Large Print 630.9747 K496g
Kimb
Good
W9-CRR-005
WITHDRAWN

May 19, 2020

GOOD HUSBANDRY

Center Point
Large Print

Books are produced in the United States using U.S.-based materials

Books are printed using a revolutionary new process called THINKtech™ that lowers energy usage by 70% and increases overall quality

Books are durable and flexible because of Smyth-sewing

Paper is sourced using environmentally responsible foresting methods and the paper is acid-free

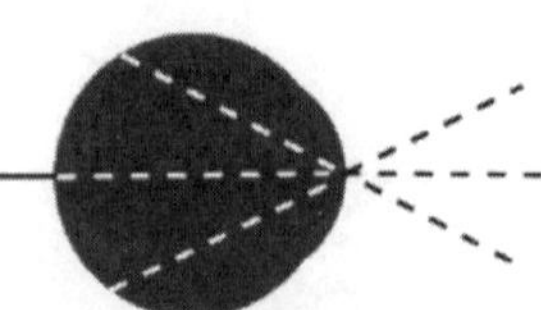

This Large Print Book carries the Seal of Approval of N.A.V.H.

GOOD HUSBANDRY

A MEMOIR

Growing Food, Love, and Family on Essex Farm

KRISTIN KIMBALL

CENTER POINT LARGE PRINT
THORNDIKE, MAINE

This Center Point Large Print edition
is published in the year 2020 by arrangement with
Scribner, a division of Simon & Schuster, Inc.

Copyright © 2019 by Kristin Kimball.

All rights reserved.

Photo Credits: page 7 © Alvin Reiner; page 47 © Nancie
Battaglia; page 141 © Megan Young Wiese;
page 395 © Tara Derr Darling

The text of this Large Print edition is unabridged.
In other aspects, this book may vary
from the original edition.
Printed in the United States of America
on permanent paper.
Set in 16-point Times New Roman type.

ISBN: 978-1-64358-525-3

The Library of Congress has cataloged this record
under Library of Congress Control Number: 2019954775

For Kelly

PROLOGUE

Some people say farming is the most wholesome job in the universe. I say having a farm is more like having a gambling problem. A farm is a living, breathing slot machine, doling out just enough reward at exactly the right moment to keep you coming back for more. Farmers are professional hopers, wafting supplication toward ancient gods we don't believe in. We hope in order to ward off disease and accident; or for rain, but not too much; or no rain, but not for too long. If frost threatens early in the fall or late in the spring, we double down on hope, trying to generate with it the degree or two of heat that is the difference between death and another few weeks of life, between a good harvest and none at all. When we win these bets, it's magic, like making something out of nothing. When we lose, the chips that get swept off the table can be measured in both love and money.

I figured this out one planting season. The fields were so dry, the wind picked up topsoil and spun it into devils that danced around us as we dug our hands beneath the surface, feeling for moisture. My husband, Mark, and I had built a new greenhouse at the end of winter, and there were tens of thousands of plants inside it—winter squash, cucumbers, brussels sprouts, and tomatoes, straining against the limit of their blocks of soil. It was time for transplanting, but the ground was too dry. The young roots that had developed under the ideal conditions of the greenhouse, with just enough moisture, just enough heat, would confront the real world of the field, which had dried to inhospitable dust. The tender roots would be sucked dry, and the young healthy plants would wilt and die. We had to wait for rain to come and water the soil.

A lot of work had been put into those little plants already. Each one started as a single seed, placed from palms with fingers into two-inch blocks of black potting soil which had been formed by hand with a special tool—a spring-loaded metal form, on the end of a long handle that was slammed into a pile of damp soil, to make compacted bricks of fertile earth. The rectangle of bricks was then released, gently, into a wooden tray. When the moisture of the soil and the force of the tool were just right, the blocks would hold their shape, with a small dent on the top to hold the seed. The seeds got a sprinkle of soil, and a watering, which would begin to wake them from their sleep.

We'd carried the heavy trays to the germination chamber, an old walk-in cooler scavenged from a restaurant that was going out of business. The chamber had a heater in it, to maintain the warm, even temperature that seeds like best in order to germinate. The flats were stacked one on top of another, six high, on open shelves. It was dark inside the chamber, and humid, and when you opened the bulky door, it smelled like the loamy floor of a forest on a hot summer day. You had to pry the flats apart and shine a headlamp across each of them to see if they had sprouted, and as soon as a few minuscule sprouts were visible, out it came, quickly, to the greenhouse. Most seeds don't need light. They carry enough energy to propel their roots down and the first leaves up. But then the energy of the seed is spent. If they are left in the dark chamber, they start searching for life-giving light, speeding up toward a nonexistent sun. In a matter of hours after germination, they will etiolate into pale, spindly, weak things that might be coaxed to live for a while by a dose of light but will never make good, and will have to be thrown into the compost pile.

Once in the greenhouse, the seedlings were painstakingly cared for every day for weeks, their soil kept moist, the temperature moderate. We woke up on cold nights to make sure the propane heaters were burning, and came rushing in from the barn on sunny afternoons to open vents and roll up the plastic walls. All this care and diligence was spent, weeks of work and many tons of expensive materials, because these plants represented a good part of a year's worth of food for two hundred people who counted on us to grow it for them.

We started Essex Farm from scratch together, not long after we met, a year before we got married. Mark was at the very edge of this wave we are still riding, made up of ambitious young farmers who grow food to sell directly to a market that is interested in how and where it comes from.

When I met him, I was thirty-one and working as a freelance writer in New York City. He was running a vegetable farm in Pennsylvania. We were not an obvious match, but we fell for each other, clicked together like a pair of magnets. He left the farm in Pennsylvania, I left my apartment in the East Village. We moved to a neglected five hundred acres in the Adirondack Park, on the rural northeastern edge of New York State, and dug in.

The farm we built was a sprawling, diversified, bewitchingly beautiful thing, composed of innumerable living parts, sometimes working in perfect synergy, sometimes descending into chaos. We created it around our desire to produce all the food we needed for an interesting, healthy, delicious diet, and to try to do it sustainably, all from one piece of land that we'd come to know as intimately as we'd come to know each other. It was a radical undertaking. That we thought creating such a diverse farm was possible was a matter, on Mark's part, of ambition and extreme optimism; and on mine, at that time, of ignorance and inexperience. I didn't know enough about farming to be afraid of it.

Our business idea was unique. "It's either brilliant or very, very stupid," Mark had said brightly when we opened our first bank account together. "I guess we're about to find out," I said. It was difficult to lift the farm off the ground, and once it was flying, it was often wobbly and not easy to steer. But it had flown. We sold memberships that were meant to allow people to eat the way farmers do—or the way they did two generations ago: a whole diet, year-round, unprocessed, in rhythm with the seasons, from a specific piece of land, with a sense of both reverence and abundance. After the membership was paid for, no money changed hands. The members were like extended family and could take whatever they wanted, in any quantity or combination, paying attention only to their own appetites and desires. In the fall, they could take extra produce, to freeze or can for winter. From our first year, we raised beef, pork, chicken, eggs, vegetables, fruits, dairy, grains and flours, and, eventually, provided some extras like sauerkraut, jam, maple syrup, and soap. In addition to our members, we supplied a few wholesale customers, a food bank, a school cafeteria, and we were still growing. What we tried to deliver, beyond food, was the same feeling I experienced when I fell in love with Mark when I went to interview him on his farm in Pennsylvania. The feeling of floating in the generosity of the earth and sun, of plenty.

Transplant is a traumatic event for a plant, even in ideal circumstances. What you want is a healthy wet plant going into a nice wet hole. Transplanting into hot dust is vegicide.

The days got longer, and the plants in the greenhouse began to look like they could use a coffee. Then like they needed an antidepressant. We kept watering them, and watching the weather report, which was full of bright happy sun icons for as far as the eye could see. Through the end of

May, we watered, watched, and hoped. Patience, Mark said. Rain will come. But it didn't.

On the last day of the month, I opened the greenhouse at dawn to find the plants had almost given up. Their leaves were going yellow at the edges. I pulled a tomato seedling from a flat, and its soil block was white with packed, tangled roots, too much young life straining to grow and not enough to feed it.

I took the plant and went to find Mark, who was welding a broken ball hitch in the machine shop. Our two-year-old daughter, Jane, was sitting on the floor of the shop, smudged with greasy dirt, tapping eight-penny nails into a length of scrap wood with an adult-size hammer. Mark scooped Jane up with one hand and took the sad tomato plant in the other. "What about a wet plant in a dry hole?" I asked. "Well," he said, "sometimes that's the best you can do." We agreed to start transplanting them the next day. The weather report offered some hope, in the form of a 20 percent chance of an afternoon thunderstorm. Not great odds, but we had to roll the dice.

The next morning I watered the seedlings before dawn and then watered them again after the sun was up, moving the water wand over the flats slowly, steadily, first one way and then the other, being sure not to stop at the edges, so they would all be completely and evenly saturated. Then I woke Jane up and took her to our friend Ronnie's house, where she was to spend the morning while I planted. I threw open the greenhouse and began to load a wagon with flats of anemic seedlings. I was four months pregnant with our second child, and the edge of the heavy flats met the first visible swell of my belly. Mark was already in the field with our young team of horses, Jake and Abby, hitched to the cultivator, drawing straight lines down the middle of each row with a dibble wheel, which left impressions in the soil every twelve inches, a template for our seedlings.

Racey joined me as soon as chores were finished. She was one of four people working for us that spring. She has a big heart and a sharp and curious mind, with a master's in nutrition, and another in international development. Racey had spent the last ten years working in Africa, first as a Peace Corps volunteer in Mauritania, then in Mali, as a consultant for the World Bank. Her father and stepmother lived down the road from us. The previous summer, when she was home on break from Mali, she had come over to help us weed our potato field and had discovered in herself the same atavistic love for working the land that I had found when I'd met Mark. After that Racey had rearranged her life, determined to acquire farm skills. Her plan was to spend the next few growing seasons farming with us, and the winters working for development agencies and NGOs in Africa.

All morning Racey and I moved slowly up and down the dusty rows. She walked along holding a flat against her hip, freeing a seedling from its neighbors with one hand, then dropping it to the ground, aiming for a dibble mark. I worked on hands and knees, scooping dirt from the bottom of the dibble, pressing the plant firmly into it. I was trying to make contact between the roots of the plant and any moisture I could find at the bottom of the dry holes. There was no coddling those babies. No plant got more than three seconds of love. Survive, we said with our rough hands, then moved along to the next. Transplanting is hard work, but it is not bad work, and it certainly isn't lonely work. While our hands were in the hot dirt, our talk ranged all over the world, from the crazy unintended consequences of international aid to Racey's international love life.

At noon, we took a break. Jane came home from Ronnie's and sat on my lap as we all ate lunch together, and then I put her down for a nap in her crib. She didn't need a crib anymore, but she still liked it and could climb out on her own. We had a system worked out, inspired by a farm family we knew whose four kids were now grown. If Jane woke up in the dark of early morning and I wasn't in the house, she'd climb up on a chair and flip her bedroom light on, so I could see it from the field and come meet her. Soon enough she would have to give up the crib for the baby who would arrive in the fall. Farm children, even very young ones, get used to relinquishing their own desires for the greater good.

Back in the field, the afternoon sun was hammering down from a cloudless sky, with an intensity particular to the North County in planting season—its heat coming through air that feels too thin and still carries the barest trace of winter. The soil looked vulnerable with its sparse covering. We looked vulnerable too, winter-white under a sun that was getting stronger every day.

We finished transplanting in the early evening. Looking back at our work was not an uplifting exercise. The plants looked like an army of little green soldiers that had dropped dead in formation. Every one was flat on the ground, stem limp, leaves wilted. The brussels sprouts were actually crisped.

While Mark gave Jane a bath, I opened the refrigerator, looking for something for us to eat for dinner. The refrigerator was smudged with dirt inside, as usual, but unusually bare except for yogurt, half a quart of buttermilk, two eggs, and a grizzled rutabaga roughly the size of my head. I pulled everything out and set it on the kitchen table, along with a pint of lard from the pantry.

Early on, when we were starting the farm, Mark and I found ourselves on nights like these driving to the gas station store, four miles north. We'd be exhausted and hungry, and making something from our own homegrown food seemed too difficult. But the chips or frozen pizza and gallon of orange juice we grabbed from the shelves made us feel worse than being hungry, full but not satisfied. Besides, we knew it was ridiculous to work so hard at growing food and then not even eat it. So we'd made a pact to always cook what we grew, even when we were busy or tired. We weren't shy about buying things we liked and couldn't grow. I would not want to cook for long without olive oil and lemons, or live—at all—without coffee. But our daily sustenance came from the farm and varied by the season. It meant that most of our free time as a family was spent in the kitchen, around the stove or the table. And it meant that Jane had teethed on braised oxtails and kohlrabi sticks, and looked forward to raspberry season like other kids look forward to Christmas. Her favorite food that spring was chicken liver pâté on toast with a sprinkle of chives on top, and she treasured the memory of a meal of breaded bull testicles that we'd made together after we slaughtered a Jersey bull the previous spring. Still, this season—the long gap at the end of winter, between the return of the light and arrival of the heat—was the trickiest to navigate in the kitchen. This was when you had to get creative.

Rutabaga is the great hulking wallflower of the vegetable world, hanging around the edge of the root cellar long after the others have been chosen. What it had going for it was staying power. Last fall's carrots were getting limp, the onions were sprouting and soft, but the rutabaga was as hard as the day it had come out of the ground. I'd hauled it upstairs a few days earlier but hadn't yet gathered the courage to butcher the enormous thing. That's the way it is with rutabaga, which means you don't eat it until you must. Farming itself is a long series of musts. The good part is that the musts you dread often end in pleasure.

I grabbed my favorite knife from the magnetic rack next to the stove, a soft steel beauty that Mark had owned before we were married. It was intimidatingly huge when we met but had gotten smaller every year from constant honing. It was always sharp, though, and knife enough to quarter the giant root. It had been so long since I'd cooked a rutabaga, I'd forgotten about its interior color, a creamy shade of mango, and its gentle cruciferous smell. Now what to do with it? I could chunk it, boil it, and mash it like potato with a splash of cream and a sprinkle of nutmeg, but we were too hungry to wait for water to boil, and I was out of cream. So I pared away the gnarly skin, found the box grater, and started shredding, which put me in mind of something like latkes. I added the eggs, some buttermilk, a few handfuls of flour, salt, and baking soda, and then ran outside to the herb garden to cut a big bunch of chives that I chopped into a fragrant pile and mixed in. The smell of the chives made me reach for the soy sauce, and a handful of green garlic, the first of the year, and a hunk of the fresh gingerroot I kept in the freezer. The meal seemed to veer away from Odessa then, and toward Osaka, settling somewhere in the middle: rutabaga-chive pancakes with soy-ginger dipping sauce. We still needed something green. There was some young lettuce in the field, and asparagus, but they were a quarter-mile dash from the house, and I was too tired to fetch them.

While the latkes sizzled on the stove, I ran back outside and clipped some dandelion greens from the base of the sour cherry tree in the front yard. The tree was as old as Jane, planted in honor of her birth. By the time Mark came downstairs with Jane in her pajamas, the orange-tinted pancakes were piled on plates, with a bright side of bracingly bitter dandelion greens that I'd laid in the hot skillet to wilt, along with a clove of crushed garlic, and dressed with a tart vinaigrette. We ate the pancakes with our hands, hungrily. For dessert, we had bowls of yogurt from the rich milk of our Jersey cows, topped with a spoonful of maple syrup, which we'd boiled down at the end of winter from the sap of the maple trees on the hill just west of the kitchen window. Jane got the thick yellow cream from the top of the yogurt, a farm-kid treat.

We put her to bed and then went to bed ourselves. We were four weeks from the summer solstice, and it was still light outside. There were no clouds to be seen. If it didn't rain, all those plants and the hundreds of hours of work they represented would die, and we'd have to figure out some way to replace them in order to feed our members. I read for a while, to try to forestall the worry. Mark, next to me, was making notes for the next day. He is missing the gene for anxiety. Unlike me, he does not spend energy considering the full rainbow of disaster that could take place. Nor does he stew over past decisions or regret stupid things he once said at a party, both sports at which I am a champion. He does not miss people who are no longer in his life. He does not think much about the past at all. But he does think about the future. Hope is a future thing, worry's ebullient cousin. He does hope. I have never met a farmer who doesn't.

I could feel the history of the day in my body. I had worn a hole in the knee of my thick canvas pants, and the skin there was pink and abraded. My fingernails felt pried up from their beds and tender from scratching in the dry dirt, and my left arm and hip were bruised from carrying the hard, heavy flats. I turned on my side to shift the weight of the baby away from my spine, closed my eyes, and watched the increasingly abstract patterns of the field replay themselves on my eyelids. As I drifted off, my hope for the plants was curled up very small. And then, hours later, I was raised out of deep sleep by the purr of distant thunder.

There was one loud clap, and Mark was awake too, and we got out of bed and ran outside together. We stood in the driveway and looked up. There was a nearly full moon in the clear sky overhead. A cool breeze was rising. We could see thick thunderheads bunching to the north and west of us, backlit by strobes of lightning. We couldn't tell which way they were moving. If the storm hit us, the plants would live. If it missed us, they would die. Odds looked somewhat worse than even. The clouds rolled and churned like breakers on a reef. Mosquitoes gathered around us, landing on our bare legs. It was past midnight, the next full workday just a few hours away, but we could no more look away from the sky than one could leave the table while the dice are in the air. It was not possible to draw the rain to us with our thoughts, but it was impossible not to try. We stood there in our underwear, moonlit, barefoot, mosquito-bitten, with arms intertwined, and hoped.

The bright forks of lightning and banks of clouds seemed to split overhead and prepare to go around us. And then I felt one fat cool drop land on my foot. Silence. Then another. Another. Then drops were everywhere, raising puffs of dust around us and drumming on the tin roof of the garage until water poured off of it, turning the slope of the driveway into a web of rivulets that joined at the base into a small stream. Mark and I held hands and grinned at each other, and raised our faces to the rain, a vestige of a dance of praise to deities we don't believe in. We stayed outside until our hair was soaked and we were cold. Not even half a mile away, just on the other side of the road, it did not rain at all that night. But on our fields, it rained seven-tenths of an inch, enough to water and resuscitate the suffering plants, all except for the toasted brussels sprouts, which were too far gone to save.

That spring was our seventh. We had a burgeoning business and some money in the bank—just enough to pay the bills and feel periodically optimistic. We had signed a mortgage to buy the farmhouse along with eighty acres of surrounding land, some of it very good land, and we owned our livestock, our horses, and our equipment. We had plans to buy another four hundred acres of land around the farm when we could afford it, and we held leases on eight hundred acres of pasture and hay ground to our south and west. The food in our kitchen, and in our members' kitchens, was bountiful and delicious, an anchor for the whirl of a busy life. We had a child and another on the way. I was thirty-nine—an elderly multigravida, in the charming language of obstetrics—and I knew how lucky that made me. It felt like our dreams and wishes, the things we spent the previous years working so hard together to tend, were coming ripe all at once.

Seven is a number with mystical appeal. A prime. A good roll in craps. It is the number of years it is supposed to take for every cell in our bodies to turn over and renew. This is not literally true—I looked it up—but it was a good metaphor for how I felt that year.. My city self, my old identity, had been sanded away by the daily work, shed cell by cell, and folded back into the soil. My new self was made of the atoms of this place, of its soil, water, sun, and dust, the way Adam came, whole, from the mud of his garden. The farm was my home, my office, my playground. We were so deep in it, we didn't know where we ended and where the farm began. We had not yet learned to separate our own well-being and that of our marriage from the well-being of the crops and animals. There is no problem with that as long as things are going well, as long as you and the farm are rolling sevens. The trouble comes when the dice go cold.

It is not always easy to see where you are from the ground. There could be a meadow coming, or a cliff. But that year, everything was good.

PART 1

ROOTS

CHAPTER 1

I had spent my twenties in New York City, in a small apartment in the East Village, working in publishing and as a freelance writer. I had never grown a thing and had no particular interest in where my food came from. That changed one summer night when, over drinks, a friend of mine mentioned a young woman he'd met upstate. She had recently graduated from Vassar and was living for the summer in a tent, growing half an acre of vegetables, which she sold directly to local restaurants. I stirred my drink, listened to the ice clink, felt the sweat run down the side of the glass in the muggy evening air. I let my mind construct an image of that tent-living dirt-farming Vassar person—she came to me as a sort of dreadlocked, feral preppie—and it was strange and interesting enough to make me want to know more. Was she the only one?

I investigated. She was not. That summer, I found, there was a string of tiny farms popping up along the Hudson Valley, started by young people with no background in farming, freshly graduated from liberal arts institutions, the ink not yet dry on their diplomas. They were creating new markets for the food that they grew, selling mostly to like-minded chefs who cared about provenance, and directly to a burgeoning sector of consumers who wanted a more authentic connection to the food they were putting on their tables. Two things were immediately clear: the work was hard, and the farmers were happy. The question that remained was: what could possibly be so good about farming to make a person willing to live in a tent in order to do it? Conventional wisdom at the time said that those with the means to get out of farming should do so as quickly as possible. Here was a small but growing trickle going the other way, and my freelancer instinct told me to chase it down.

I'd soon chased it to central Pennsylvania, to a young farmer I kept hearing about, a lanky, loquacious, sharply intelligent, and ridiculously energetic man named Mark. He was a first-generation farmer too, but he had been at it for a decade, apprenticing for some of the most experienced organic growers in the Northeast before starting his own farm, on leased land. He was raising vegetables when I met him, plus some beef, pork, and apples, for a hundred families who signed up for a share of his produce. His single-minded enthusiasm for growing things was virulently contagious. He was an amazing salesman for his food, and also for his vocation, and he was making career farmers out of some of the unsuspecting young people who had come to work for him in what they thought was just a summer job.

I was thirty-one that year, single, and longing for some things I could barely bring myself to name. Those longings found a surprising answer in Mark, and in the work he loved—that alchemical combination of sun, earth, and physical effort. My interest in him and in farming quickly shifted from professional to personal. The shift was exhilarating, terrifying, and irreversible. Taking it back would have been like taking back a lightning strike or a plunge over a waterfall: impossible, even if sometimes I would have liked to try.

When I fell in love with him, I'd been living in the city for almost ten years. I'd traveled a lot, to interesting corners of the world, unhindered, which made me feel alive. I told my sister I thought this man might be my greatest adventure. She took in the larger picture—the dirt, the blood, the work—and asked if she needed to take me to India for a few weeks to help me clear my head. This was the course of action she and I had agreed on years before if one of us saw the other about to make a dubious decision. "No," I said, "I'm pretty sure."

In less than a year Mark and I started a new life together on five hundred acres of good soil in Essex, New York. The next fall, at the end of our first growing season, we got married in the loft of our barn.

From the time he was twenty and discovered his calling, Mark's inner radio has been tuned to WFRM, all farming, all the time. I don't know anyone more dedicated to anything than Mark is to farming. The farm is his focus, his daily work, and his identity. The word "farmer" makes you think, perhaps, of qualities like sober, modest, plain. Also, I bet, a man of few words. But in order to understand the first thing about our farm, in order to understand our family, and I suppose in order to understand me, you need to grasp what my husband is like. Imagine the energy of a surprise party thrown for a band of uninhibited monkeys. Distill that energy and bottle it into a lean, strong, six-feet-six-inch form. Add a heavy shot of brainpower and stop it with a cork. Now shake hard until the cork blows off and it all bubbles over. That is Mark. When he is not talking, he's moving: drumming, juggling, chopping, or sometimes shadowboxing at your face. When he has a repetitive job to do—picking peas, say, or sharpening knives—he has a way of pausing first to center his weight over his hips, his body calculating the pattern of least resistance, and then launching at full speed. But usually, he is talking, with words as fast and big and forceful as his hands. He enters a room talking. He talks to your retreating back. Sometimes, if you happen to talk, he interrupts. He doesn't mean to be rude, but neither does he care very much that he is being rude. Talking is his way to figure out what he thinks. He needs it. And he needs, above all, to connect, to be seen.

During college, on April Fools' eve, he led a daring (and, over the years, growing) cohort in acts of campus disruption. The first year it was predictable stuff, like rearranging the university president's office furniture on the lawn. But as each year passed, the pranks became bigger, more imaginative. By the fourth year, the group numbered in the hundreds, and the keystone prank was the midnight relocation of his college's prized Calder sculpture, a priceless behemoth that they hoisted to the top of the tallest building on campus. They created cover for themselves by putting up posters all over town, advertising a purely fictional event: "A Multicultural Perspective on Modeling in America," a lecture by Fabio.

The memory of those nights still makes his face light up. I'd heard about the escapades for years, but it took me a long time to work out the parallels between that time and our life on the farm. The pranks always included enormous challenges, an element of physical danger, a cavalry of willing workers, and a transgression of tacit rules. Mark was the charismatic leader of a large band of subversives. As more people came to work for us, the farm began to take on a shape that resembled a perpetual replay of April Fools' eve: a lot of people doing radical things under his direction, quickly. This was sometimes a lot of fun and at other times incredibly stressful.

Mark is as strict with his own rules as he was loose with others'. His favorite philosopher is Marcus Aurelius. Peel away the outer layers of jester and hippie, and what is left is a powerful stoic. Two or three times a year, he begins a new self-improvement habit. One winter, he read every historical play in the *Yale Shakespeare*. He took up meditating and, unlike me, stayed with it; he hasn't missed a day in over six years. He fasts twice a week, runs wind sprints at dawn on Mondays, and ends every shower with a sixty-second blast of ice-cold water, because he read that those things are good for his mitochondria. He is left-handed but believes there is virtue in being physically balanced, so he forces himself to use his right hand for daily tasks exactly half the time. He has recurring dreams about saving strangers from mortal danger. When he dreams of falling, he pulls himself out of it and makes himself soar.

I used to think that what delighted him most was feeding people. But now I think that food is the portal to something bigger. The idea of a farm is so simple: catch the sunlight, hold it where it fell, and use it to meet human need. On our farm, we ask the sun to meet as many of our needs as we can: food year-round for all of us, but also sugar from the trees, soap from our animal fats, electricity from the solar panels, wood to heat our home. The sun is a currency, the world's first and most basic, endlessly convertible to other forms. You can take it in cabbages, carrots, or corn. You can take it in grass or hay, in milk or meat or timber. Give it several million years and you can take it in oil. But managing the conversion takes effort. What makes Mark ecstatic is the work itself, which is the connection between the sun and all life. "We're sun brokers," he likes to say, which brings to my mind an image of Mark as Apollo, in French cuffs and a power tie, hustling sales of pure light. But that sense of powerful magic is what Mark runs on, and food is just the way to reveal it to people.

Once, on our way to New York City with a delivery, we stopped to pay the toll at the Tappan Zee Bridge. Our old Honda Civic was so stuffed with boxes of food, I was occupying a body-size cavity in the passenger seat, immobile. It was late, we were behind schedule, I was tired and increasingly claustrophobic. “Looks like you’re moving,” the toll collector ventured through the window as Mark dug in the cup holder for change. “No, we’re bringing some food from our farm down to the city,” he said. “Oh, you have a farm?” she asked pleasantly, which, I knew from experience, was exactly the fish he was angling for. As he began to describe it in enthusiastic detail—the acreage, the animals, the season, the harvest—cars lined up behind us. I buried my head in my hands and stuck my thumbs in my ears to block the noise of the horns. Through my fingers I saw him reach behind his seat and pull out a dozen eggs, a jar of maple syrup, a bag of spinach, ears of corn. Each item passed from our car into the tollbooth and was met with a peal of surprised laughter. When we finally pulled away, he turned to me, beaming. “She was so happy. She’s going to come visit. She’s awesome. You’ll love her.” That goes a long way toward explaining what motivates Mark and why we have so many visitors.

• • •

Our farm sits a mile from the tiny village of Essex, on the shore of Lake Champlain. Every generation before our time contained enterprising newcomers laying plans for when the wider world would discover the town's charms—the quaint ferry dock, the restored Greek Revival homes along Main Street, the view of the Green Mountains to the east, the High Peaks to the west. It is tucked into the majestic landscape of the Adirondack Park, where zoning prevents sprawl and development, and much of the land is constitutionally protected, forever wild. On this sure-seeming bet a few real estate deals are done each decade, but then the town sleeps again.

The lake divides New York from Vermont. It is also a boundary between two different economies. Our side is one of the most sparsely populated regions in the United States. Demographically, it has more in common with empty stretches of the American West than with other regions of our own state. Jobs are hard to come by. The public school district that our children attend consists of one school: a two-story brick building built in 1932, where the elementary students occupy the first floor and then, in a scripted rite of passage, move upstairs for middle and high school. The total number of students is a little more than two hundred, and the graduating class sizes range from ten to twenty-four.

Across the lake in Vermont, not three miles as the ferry runs, things are different. The town where the ferry lands is casually genteel. The clapboard houses are carefully tended and occupy plots of expensive land. The public school system is well funded. The city of Burlington, visible to the north, has a university and a teaching hospital. On our side we have a tampon factory, now closed, and a mine that brings out most of the world's supply of an asbestos-like mineral called wollastonite. When the sun sinks into the mountains to our west, our side goes dark, and the lights of Burlington shine like new coins along the eastern shore.

The year-round population of our village at the last census was less than seven hundred and has been in a steady decline since a shipbuilding boom in the 1850s. Nearly all of the houses on Main Street were built during those years. Most of them are restored, and many belong to people who live in them for only a few weeks in summer. In winter, they stand dark and empty, the stores and restaurants on Main Street closed.

When I moved here from Manhattan, the ferry, and its promised connection to the city on the other side, made the transition seem possible. But we had a string of cold winters in our first years, the lake froze hard, the ferry stopped running, and our tie to the larger world on the other side was cut. In the hours not filled with farmwork, we went skiing or skating and kept the farm pond clear for hockey games that ended in early evening, with headlights pointed in. By the time the weather pattern shifted and the ferry ran all winter, the lights across the lake held no appeal. We crossed once or twice per year, but our lives were firmly planted on our side.

In summer, people arrive to occupy their houses, bringing motorboats and money to spend. The restaurants put out their signboards, the ice cream shop hauls the chairs and tables to the sidewalk, and the antique store puts new price tags on the shabby-chic merchandise. There is a palpable feeling of urgency every Memorial Day weekend. There is hustle. The financial success or failure of the year, for most people, hinges on what can be made before Labor Day. From our fields we watch the pulse of traffic go by as the ferry deposits summer families and fresh visitors.

There's always been room in Essex for a few eccentrics, especially friendly ones, but it's not easy to jam a big personality like Mark's into a very small town. When we arrived, people generally decided he meant well, and accepted him. Sometimes, though, Mark went too far. He was too tall, too pushy, or far too immodest. In his own way, about farming, he was too ambitious. It's not in his nature to notice social cues, but people let me know when a line had been crossed, through second- or thirdhand comments.

"Your husband . . ." they began. And I'd brace myself for what was coming next. It might be about the crooked row of rusty equipment he left parked at the end of the driveway, where it was convenient for him to hitch to when he needed it but an eyesore for everyone who passed. Or the trio of slaughtered pigs, open-necked and dripping blood, swinging by their hind legs from the bucket of the tractor that he drove through the village at the height of tourist season because it was the most direct route between the pasture where he'd killed them, and the butcher shop in our farmyard, where he would gut and skin them.

It wasn't always easy for me to fit in, either. There was a divide between the summer people and the locals, whose roots went back generations. It was also a division between the haves and have-nots, between liberals and conservatives, between those with fancy educations and those with common sense. Also between the people who worked with their hands and the people who paid them for that work. Despite the small population, they didn't usually attend the same parties. Town couldn't always decide which side I fell on, with all the mixed signals I was giving off, a newcomer with city manners but scarred-up, callused hands.

When we first arrived, I was always lost, even though there are not many roads. "How do I get from here to there?" I'd ask. "Oh, you go past

Lincoln's, just before the old Ansen place, turn at the Four Corners, and they live up Bull Run." These were signifiers without actual signs, not marked on a map, their meaning attached to our town's collective memory. That system assumes a deep familiarity with the whole community—not only its buildings and geography but its history and its people. To know where Lincoln's was, you would need to know that the empty steel building west of our farm used to be a feed and hardware store owned by the much-loved, kindhearted, and long-deceased Bob Lincoln. To know Bull Run required understanding that the rise past the south end of Main Street was once the domain of a family of cantankerous brothers. The name referred to their belligerent natures, still in play a hundred and fifty years later. I got this explanation from Jim LaForest, one of our town elders, who remembered hearing it from his father-in-law, who knew the brothers not in his lifetime but through family lore. To know that our hill was called Rogers Hill presupposed knowledge that a farmhouse owned by the Rodgers family used to be at the top, where an empty restaurant stands now. Anyone who'd seen that house with their own eyes, or known the family, was long dead, but the name had stuck to the place.

It was the same with people. Our first winter on the farm, our neighbor Ron came over, bearing a venison pie he had made, its rich meat filling fragrant with warm spices, a perfect ratio of gravy to flaky butter crust—the first of many satisfying delicacies he brought us, and the start of countless kitchen conversations. "Aunt Shirley likes meat pie with a touch of nutmeg, but Danny prefers it without," he said without preamble. I had no idea who Aunt Shirley was, nor who Danny was, but Ron kept going, dropping more names with every sentence. I decided not to stop him to ask for clarification. I already knew I wanted to belong, and figured it was better to learn these things like a foreign language, by immersion. And eventually, I did. As we wove our lives into the fabric of the community, I learned that Aunt Shirley was Mrs. Shirley LaForest, Jim's wife, who taught Sunday school at the Methodist church. She was the official town historian, had grown up on a farm nearby, and could tell the history of the local farms and what each family had grown or raised, traced back a hundred years. She knew which farms had specialized in growing trefoil for seed, and which had raised sheep, and which had been dairies shipping butter on the train that cut over the road just west of our farm. All three were agricultural specialties in our region at different times. Now Shirley was watching a new farm grow up on our land, just one in a line of many.

Danny turned out to be Danny Sweat, our

mailman, who was also Ron's best friend, an EMT, and a commissioner for the fire department one town north of ours. He was Ron's business partner in his sugarhouse and one of his hunting companions. Ron himself was rooted to the town by the weight of nine generations, the branches of his family spread wide through the region. He was a widower, retired from the merchant marine; he had an engineer's logical mind and a deep love for this place. He also had a sense of duty to the whole community. He'd served at different times as the chief of the volunteer fire department, the town supervisor, and always a volunteer EMT, pulling out of his driveway at all hours in any weather to fetch the ambulance from the firehouse and tend to a neighbor's crisis. The complexity of all the lines of connection took me years to untwist. Following them all the way to their ends gave me a lot of satisfaction.

Mark and I came from different backgrounds, had different values, different motivations, and a different perception of the place we'd landed in together, but the farm was the sturdy bridge between us. During the first years of our marriage, we worked sixteen hours a day together to build it. What it gave us in return was the scaffolding for our relationship, for a family, a life.

And it gave us food. In my old life, food was usually not much more than a need that had to be met every day. Here, it was the center of everything, who we were and what we did. The quality and taste of what we grew was so different from what I had been used to that I felt like I'd discovered a lost continent. Seven years in, I still couldn't get over it. Seasonality had been invisible to me in the city, where anything was available at any time. Here, every week brought new flavors to perfection, and then they were gone, replaced by others. It was a never-ending cycle of longing and fulfillment, directly connected to our work, and learning to live this way was like hearing a tune I had known once and forgotten. It just felt deeply right. After Jane was born, my happiest moments were in the kitchen, sharp knife in my hand, small child underfoot, taking apart something that, hours earlier, had been warmed by the same sun that was warming me, and fed by the soil under my feet.

All farms reveal your inner self whether you like it or not. Your daily choices shape the soil, fields, buildings, and fence lines, sketch the plants and animals in your own soul's likeness. Your values are visible in the way a farm looks. Your ambitions, strengths, and weaknesses are there for everyone to see. The farm we made was a physical manifestation of who we were, together.

When we'd arrived, with a measly fifteen

thousand dollars in savings between us, the land had been out of regular production for almost two decades. The soil hadn't been asked for much in that time, which meant it was in pretty good shape. Bringing soil back to health after it has been misused can take years, sometimes generations. We had all sorts of soil types, from gray impenetrable clay to light sand. Millions of years ago, this land was under a vast shallow ocean, and the sea creatures laid a bedrock of limestone, which made the earth sweet. Then came ice sheets a mile thick, pushing down from the north, melting back, leaving rubble and erratic boulders in their wake. The meltwater rushed across our land and filled the basin of Lake Champlain. The lake covered most of the farm, but its edge is now a mile from our house. A stretch from our house to the neighbor's was a freshwater beach. Clams grew there in abundance, and when the land dried, topsoil formed over them. Occasionally, we find their shells, a half-inch long, bleached white by twelve thousand years underground. They aren't fossils but actual shells. Mark considers them lucky, and when he finds them, he eats them, crunching them between his teeth. The very best acres were formed when the lake level was six feet above our fields and primordial storm surges roiled the waters. The turbulence dispersed the fine silt that forms clay and left behind the coarser particles so that the soil is not as dense as it is in other places in our valley and is more hospitable to the roots of plants. We have more than a hundred acres of the good stuff in two different places.

The soil was a gift. The rest of it—sagging fences, leaking roofs—was fixable. We spent much of our first year pulling down what couldn't be saved and patching what could. Over the ensuing seasons, we added enough infrastructure to qualify as a small nation-state: three new barns, a new well, improved roads. Six years in, we dug the footings for a grant-funded solar array to supply all the electricity our farm required then. The solar project was part of our plan to reduce our reliance on fossil fuel. We wanted to try to make a farm that did more good than harm to the soil, the water, the climate, and the community, as well as be productive and profitable, which is much harder than it may seem from the outside.

Mark and I had long talks about this in the beginning. It looked fairly straightforward to me—feed people, be nice, don't wreck the land—but the longer I farmed, the more complicated it became. Were we adding health to the soil? Producing what the land could reasonably carry? Protecting the quality of the groundwater? Were we paying our workers a living wage and treating everyone fairly? Were we sequestering more carbon than we released? Were we keeping the farm financially stable? Did we have enough time away from farming to be a healthy couple, a healthy family? Any one of those things alone was a straightforward proposition. The challenge was to do them all simultaneously. Mark wasn't convinced there was ever a time in the ten-thousand-year history of agriculture when farms added more, on balance, than they took. In his obsession with farming, he worked to figure out if it was possible.

The draft horses were part of the plan, and that was where his obsession and mine intersected most neatly. We'd built the farm around them and the work we could do with them. We had tractors and used them, but we relied on the horses for most of our vegetable work, more than half the haymaking, and the hauling of heavy things that is a constant part of farming. Their presence had shaped the farm in the imperfect look of our furrows, the shaggy headlands where the horses grazed at night, and the small scale of our fields, which ran long and narrow to minimize the time spent turning a team around. For Mark, the horses were an expression of his desire to use the sun in real time: the energy of the sun created the plants that fueled the horses that did the heavy work with us to tend the plants that fed the animals that fed us. For me, the horses were a rediscovery of my childhood, when my whole world had revolved around them.

CHAPTER 2

The roots of my love for horses go as deep as memory. The first: We were visiting my mother's family in Tennessee in 1975. I was four. We were Yankees and they were Southern. Mom was somewhere in between, assimilated Yankee. She'd keep her rural Southernness well hidden until those annual trips; it had been shamed out of her by the girls she'd roomed with in New York City, when she was a beautiful nineteen-year-old stewardess for TWA, freshly plucked. They would make her repeat things and then laugh at her accent. But on that trip home, with us children, without Dad, I heard her vowels change back to their original shapes; her voice and manners became softer, more delicate.

My brother and I had been coaxed into good behavior on the airplane to Tennessee with the promise of visiting a pony that lived at my uncle's house. It had sounded like fantasy or maybe a metaphor. As if my mother had told me that my uncle had a benevolent dragon living in his yard and we'd get to see it if we were good. At his house, I became aware that this pony was real. The grown-ups sat around the living room, talking loosely of taking us children outside to ride her. I was full of the small powerless person's fear that it might not come true, that my great longing to touch the pony might continue to go unfulfilled because of some grown-up snag. I could see that my cousins, my brother, none of them felt the way I did. To my cousins, who lived there, the pony was a chore.

We pulled up to the barn in my uncle's Cadillac. The Tennessee landscape was a mix of red clay and brown grass. The barn was metal-sided, long. The adults made small talk as we walked in, but I, the family chatterbox, was suddenly hushed. I'd already caught a hint of it, in noises of the barn—the blowing, stomping, shaking of large bodies. The mysterious new smells of sawdust, leather, the horses themselves. Someone lifted me up. I saw an enormous eye. I put my hand between the bars of a box stall and touched the velvet lips of a beast. There was a gust of warm breath from nostrils as large as my fist, and I felt an unnameable physical thrill. I had no word for it then. Now I'd call it awe.

I already knew that animals were more interesting to me than anything else. Partly because of their complexity, partly because they were unattainable. For an afternoon, I'd owned a painted turtle that someone gave me, but his cardboard box blew away with him in a thunderstorm that same day. Once I'd held a spent butterfly with tattered wings. Our old black-and-white cat tolerated me, but barely. I liked to press my face into the fishing-line feel of his whiskers, his pink tongue coated with Tender Vittles, the wet-chalk smell of the pads of his feet. He held no mystery for me, or nearly none. But no one would let me near the large or wild animals that I ached to touch. Here, for the first time, I could, I did, and I knew it was a great privilege.

Emotion is the glue that cements memories; the stickiest stuff is reserved for the deepest and strongest feelings. I have only the smudgiest recollection of the catechism I learned for my confirmation. I have lost every formula I memorized in calculus. I don't even know what I had for lunch last Wednesday. But I remember in sharp detail the saddle that my uncle pulled out of the shed and placed on the pony's back that day over forty years ago. It had tooled light-brown leather and a little silver horn. I can see the bridle going over the pony's black ears, which were fuzzy with winter fur. I stood as close as I could so that I would be first. The saddle was like a throne. The feel of reins in my hands was responsibility, power. Then movement, as my uncle led the pony in circles through the yard and down to Pretty Creek. That rocking, magical movement. I planted the pony's name, Gaye, emphatically in my memory, knowing that I'd have to survive on pretending, until I could do this again.

I don't know where my affinity for horses came from, but my uncle had it too. He had discovered it against all odds. My grandparents did not have it easy. Their years of raising children were rocky. My granny had an eighth-grade education; when she was sixteen, she went to work in a shirt factory. She was working there when she met my grandfather, who was running a restaurant in town. My grandfather had finished only sixth grade, and early on in the marriage, he drank. He would disappear on drinking binges a few times a year, until my mother turned ten and asked him to quit. He abruptly did. They moved often, worked seven days a week at service jobs or their own small businesses. For a while, they had a small motor inn with a beer joint. Granny weighed ninety pounds in her prime and was tough as flint. We went to see her every year, except the ones she came north to see us. Oh, how I loved her. I kept her stories and the sound of her voice in the same place I tucked the silver-horned saddle. Once a man wandered drunk and threatening into her house at the motor inn while she was alone with her children. Granny gave him a chance to turn around, and when he didn't take it, she grabbed her loaded pistol from her dresser drawer, held it in two trembling hands, and backed him out of the house.

They were not people who had room in their lives for something as frivolous as horse fancying. But my uncle walked into a barn full of beautiful horses when he was a young man just out of the navy and did not leave that world for the next fifty years. He became a world-class rider, then a famous trainer of champion Tennessee Walking Horses. I think he recognized what horses did to me, and sent me a subscription to the *Walking Horse Times*. By the time I was a teenager, I knew the famous Tennessee Walking Horse stallions like other girls knew the pop stars in *Tiger Beat*. I cut out newsprint pictures of the champion horses and taped them to the back of my bedroom door.

Maybe the affinity is genetic, but it skipped my mother completely. Neither she nor my dad was a horse person. My interest wasn't shared by them or even understood, but it was, when possible, indulged. They were not wealthy, but they used some of their limited resources to grant my singular wish. As a parent now, I see what that was. Pure love.

I perched next to my mother, watched her flip through the Yellow Pages for "Stables." I was seven. I listened to her arrange my first riding lesson, setting a day and a time. "And what do we do in case of inclement weather?" she asked the person on the other end of the line. I put that unfamiliar word, "inclement," into my brain, in the same place I had put Gaye's name, because if it had to do with horses, it was important.

For the next several years, the best hour of the week began at four-thirty p.m. on Thursdays. I used to pray that I wouldn't get sick on a Thursday, and for the weather to stay above 15 degrees in the winter, because when it was colder than that, the sweat would ice on the horses' chests and we could not ride.

The barn was full of horses and ponies of all shapes and sizes. Everyone rode English. We all learned to jump. The names of those long-dead horses are still alive in my head. Not the people—they are all a blur—but the horses. There was fat little Daisy, the Shetland, on whom I learned to tack up. She was temperamental but quick-footed, and she lifted me over my first crossrails, introduced me to flight. Goldie the draft cross gelding; Vicky the bay Morgan; Chips, a chestnut cob; Big Satin, Thoroughbred; and Little Satin, a fierce small black gelding saved from the kill auction and reformed into a school horse by Judy, the instructor. Judy was the exception to the blur of people at the barn. I cared very much about what Judy thought of me. She was powerful. The horses knew it and so did I. I remembered every single thing she told me, and did my best to please her. She seemed to understand that small children on horses were not cute, that for me this was serious business.

Days other than Thursdays were tolerable, because if there weren't horses, at least there were books about horses. There were instructive texts on equine diseases with antique names like poll evil; books on conformation, breeds, and the finer points of equitation. And there were narratives. *The Black Stallion*, *Misty*, *Black Beauty*. Later, *Xenophon*. I read them all.

When I turned fourteen, my parents bought me my own horse, Vicky, the bay Morgan mare from Judy's stable. We boarded her at Smith's Last Chance Farm, a barn a mile from our house. It was very much a barn, not a stable. Smith kept heifers, pigs, and sheep in there, along with my horse and his daughter's. Most of my teenage years were spent in that barn or galloping through the fields around it. The horse absorbed what I needed her to absorb, which was a combination of obsessive love and physical energy, at an age when those things can be dangerous if pointed in the wrong direction.

The summer I was eighteen, we sold her. We had to. I was leaving for college. I rode her bareback across the fields to her new owner's barn because I wouldn't part with my saddle. Then I walked home, blind from tears, the bridle over my arm.

When Mark and I started farming with horses, we picked up a thread that seemed about to be lost to the fabric of time. We hooked on to it and made a link at the last possible moment. The tractor hadn't completely displaced draft horses in this part of rural New York until the 1960s. There were men in our town who had grown up working horses on the farms all around us. Those men were native speakers of draft horse work, unlike us. We learned it late in life, so will always speak it in an awkward pidgin. There was Mr. Christian, who would come by with his sons and grandsons to give his terse opinion, only when asked, on a horse that looked a bit off, or to evaluate a new team. Mr. Christian chewed tobacco every day for seventy years. One morning he felt unwell. They operated, saw the masses of cancer that had piled around his stomach, and closed him back up. He died a few days later. Into the ground with him went the memory of the crossbred team of mares he had when he was a boy, the feeling of what it was to work them.

Then there was old Shep Shields, obscene, arthritic, and incontinent; the stories of his philandering days were legendary in our small town. He worked Belgians on his father's dairy when he was a boy, and after tractors came, he kept a team of horses on his own farm, for the love of them. There is a picture of him in our firehouse, driving a pair of snappy Belgian mares in the Fourth of July parade, pulling the firehouse's antique pumper. When he could no longer care for horses, he hauled feed, driving around the neighborhood with a load of corn silage in the bed of his truck, a six-pack of cold beer in the cab. The smell from the cab was unmistakable: urine, sweet silage, beer. He would park on the side of the road when I had horses in the field, and wave me over. On hot afternoons in the summer, he would pass a beer through the driver's window to whoever would come for it; he had to get rid of the evidence before he went home to his wife. When his wife died, he took up with a woman roughly half his age. It seemed like he would live forever, pickled by his own delight in that arrangement. But then one day even Shep was gone, and our link with the working horses of the past snapped.

Once I was working across a field with Jake and Abby on the disc harrow, knocking large clods of soil into a smooth seedbed. The work was hard, I was tired, the horses were tired. I didn't have a watch on, but when the wind had shifted, I'd heard the church bell chime noon. When the wind shifted back, I could time the day only by the ferry traffic. Every half hour, there was a stream of cars with different plates, and then the road was quiet again.

I whoaed Jake and Abby in the middle of the field to give them a rest and wait for the ferry traffic to pass. I got off the seat to stretch, hung the lines over my wrist. That was when I saw that the last car in the line from the ferry had turned around in our driveway and stopped. I got back on the seat and saw that the car door was opening. "Come up," I told the horses, ducking my head, pretending not to see. I had a long way to go and didn't have time to chat with tourists. But on my next pass, I realized three people were heading in my direction. A man, a woman, and between them, another man, very old. They carried him over a ditch, and his feet came down on the soft dirt of the field. They were walking toward me, slow.

The horses stood perfectly still, ears up, watching. I could have made it easier, should have moved a little closer. But the horses were content to stand and rest. There was sweat on their necks, and they were breathing hard. The heat of their bodies carried their smell to me, diffused it to the heavy air all around us. The three arrived.

"He grew up on a farm," the woman said. "He's ninety. He used to drive a team of horses."

"On my father's farm," the old man said.

"He just wanted to touch them," the woman said.

The old man laid his hand on Jake's neck. Jake bobbed his head and blew, shook his body, made the tug chains rattle. The man was far off in his memory of a horse that had been dead seventy years; the corners of his mouth were happy. We looked at each other. We were a couple of compatriots, different ages, different genders, different times, different horses, but with an understanding. A link.

Mark and I had eight draft horses on our farm that year, plus a fat white pony. The pony, Belle, was almost short enough to walk under the draft horses' bellies, and saucy enough to try it. She'd belonged to our friend Scott Christian, but his kids had outgrown her, and she'd been languishing for a year in a box stall at his place. Scott had come by one day the previous winter and offered to sell her. Mark, who couldn't see much benefit in having a pony to feed and care for, had declined. But Scott, who is a clever judge of people, came by again and found me at home alone. He offered her to me, this time for free. "You're going to need a pony for Jane," he said. This, I thought, was absolutely true.

It had just begun to dawn on me that being a

kid on a farm involved some trade-offs, and I would need to be aware of them in order to make the balance of childhood come out on the right side of even. For one, we were poor. There was a good chance we always would be. The biggest part of every dollar that came in went right back out and was sunk into the farm. If we ever spent money on something other than the farm, we were ridiculously frugal. We had to be, if we wanted the farm to survive. I didn't *feel* poor. We owned this beautiful and productive land, plus our house, and we ate like royalty every day. But now that we had a child who was beginning to be conscious of the world around her, I realized that even if we didn't feel poor, we usually looked it. Our car had 180,000 miles on it and was pocked with rust. We never bought new clothes, and I cut everyone's hair.

Mark had always been oblivious to markers of socioeconomic status and mostly blind to the way he appeared to others. He truly didn't care what he wore and never had. The only thing he asked of his clothes was that they sufficiently cover his nakedness, and he kept them for as long as they could fulfill that basic duty. I came from a middle-class family who put a predictable emphasis on the way people looked. When I'd arrived in Essex, I'd made an effort to stay presentable, but the farm soon knocked it out of me. There's no point in caring about clothes when they are likely to get ripped on barbed wire by the end of a week, or permanently stained by blood or manure before that. Besides, who was going to see me on the farm except Mark, who didn't care? By the time Jane came along, it seemed absurd to me to spend any money on clothes, especially when there were so many hand-me-downs circulating in our community. I don't think I bought a new piece of clothing for Jane until she went to school, and then it was just one outfit to mark the occasion. For myself, I got by on the boxes of discarded clothes that my fashionable sister sent me from the city. They were a year or two out of date in New York, which meant way ahead of the style curve in town, and I desecrated them quickly. Every six months or so, I treated myself to new socks. Once our mechanic, Jeff, had given me a long look while I waited for him to right some wrong in the belly of our car. "You know," he said, "you've changed a lot since you moved here." "In what way?" I asked, unwisely. Did I get fitter? Funnier? More skilled? "Well," he said, almost to himself as he went back to his work, "you're a mom now." I pulled up my stretch pants, wrote him a check, and drove away, disheartened.

We weren't much richer in time than we were in money. We were either working at something or asleep, which didn't leave room for the sort of child-centric things I saw other parents doing with their young children. Summers, when other families kicked back a little, we were in the thick of it. Even if we could afford it, taking Jane to Disney World for a vacation seemed as impossible as taking her to the moon. Forget a vacation. We milked cows. We couldn't be away from home longer than twelve hours, at least not both parents at the same time.

I wasn't sure yet how these forces would shape her childhood, but I was aware that this one was the only one she would get. I had chosen farming when I was a fully mature adult who had seen a lot of the world and had other choices. I'd decided, on thin evidence, that what I'd get from it equaled what I knew I was giving up—things like predictable hours, a paycheck that arrived even in the wake of a flood or a drought. I'd reckoned it was a fair trade, for me. What I couldn't have known back then was what it would feel like to have made this choice for a child too. What would *she* get as recompense for the hard parts?

Sometimes I'd make a list of the good things and tick them off on my fingers when I woke up worrying in the middle of the night. She'd have the best food. She'd have the run of five hundred acres and a sense of complete physical freedom. She'd belong to a place more deeply than most people. She'd get two parents who were very busy but always present, right here at home, working hard at what they loved and believed in. And she'd get to grow up in the intimate company of nonhuman living things, the sort of company that was entirely normal once, which I had craved as a child but which has become increasingly rare. She'd have garden spiders, barn cats, dogs, cows, pigs, crayfish, and chickens to play with, woods and fields to explore.

And she could have a pony. I thought of the first time I'd touched a horse's nose, the breathless joy that gave me. Other things I'd like to give Jane might be out of my reach, but this one wasn't. I'd never laid eyes on Scott's pony, but I quickly decided he was right. We needed a pony. So I said yes, and before Mark could come along and countermand my decision, Scott went home, hooked up his trailer, loaded the pony, and delivered her to the door of the West Barn.

In my imagination, she'd been perfect. In reality, she had overgrown hooves, a large belly, and a thick white coat full of dust and pied with mud-brown spots. When she took a deep breath, she coughed. The places around her eyes and nose looked too pink, like those of an albino rabbit. The eyes themselves were hard and canny. She had two brown ears and a brown cap between them, a pattern that Native Americans call medicine hat, which is supposed to give a horse special powers. I liked her immediately.

The first time I brushed her, pulling out clumps of white fur with a shedding blade, she turned her small butt toward me and offered to kick, a gesture that seemed almost laughable to me after seven years of working with horses that weighed a ton each. But when I put my old saddle on her back and gently tightened the girth, she laid her ears back, dove at me like a striking snake, and bit me hard on the thigh. I pulled my pants down to see if I was bleeding and found the impression of her teeth in a hard blue bruise. Mark walked into the barn and found me like that, pants down, next to a pony he didn't recognize. He took in the scene, realized what I'd done, and shook his head. The bruise was already impressive. "That pony," he said, "is going to be the most dangerous animal around here." I felt an allegiance with her despite the bite: "She's not dangerous. She's opinionated." We were two small creatures on a farm that often seemed scaled too big for us, and we had to assert our power where we could.

Belle was short but sturdily built, just big enough for me to ride. That was a lucky thing, because she was the kind of pony who believed in her own free will and had developed some ideas about what she should and should not have to do, ideas that we'd need to straighten out before she could be Jane's pony. Once I was aboard, bareback, Belle settled down and was game for anything—a trot along the firm ground between the raspberry bushes or even a good gallop through the flat, grassy alley made by two rows of linden trees. When I turned her out with the draft horses for the first time, I watched for a tense few minutes as the big horses ran her around the paddock, her ears pinned, her small legs running double time to their beat. The earth shook. I was afraid they would run her right through the fence, and I pictured myself picking up pony parts. Then the other horses lost that avid quality, slowed, circled, squealed a little, and put their heads down to eat. When one of the geldings got too close and pushy, Belle gave him a kick low in the ribs, swiftly establishing her place in the herd order.

Over the winter, we got to know each other and worked on her faults. Unlike the calm, steady draft horses, the pony had a flair for drama. Sometimes she would spook so wildly at an innocent object, I actually suspected she was acting, and not very convincingly. A raised voice or stern gesture from me in response to something like biting solicited an overreaction from her, which escalated the situation instead of correcting it. I figured out the best way to train Belle was to arrange consequences for her undesirable actions that seemed accidental. The next time I tightened her girth, as she pinned her ears and swung her head toward me to bite, I casually positioned my elbow in the path of her nose so that she collided with it. I pretended I hadn't noticed and continued about my business. She put her ears half back, thinking, and tried it again. A few such bonks, and the biting stopped.

That spring, just in time, Belle was ready for a job. I was pregnant, and when the exhaustion of the first trimester struck me, Jane was still too small to keep up when we walked around the farm but felt too big to be carried all the time. She could sit on the pony while I led her, and we could go anywhere together at a grown-up's pace. As soon as she was employed, Belle perked up considerably. She seemed to like carrying a child and began looking for me when I came to find her in the pasture, instead of running away. Jane developed good balance on that spooky pony and learned to pick herself up and get back on when she fell off.

CHAPTER 3

There's no month or week or even day when there isn't work to be done. I've never once looked around and thought, *Huh, I guess we're all caught up*. But the amplitude of the work rises in the spring. The rhythm of the day is fast and urgent, because the outcome of the entire year hinges on what happens now. All sorts of disappointments can befall a farm in other seasons. There are summer floods, early frosts, and deadly blizzards, but in spring, it's all pure potential. If spring work is left undone, you're not even in the game. You can't harvest something you never planted. The sun pours in, all that gold hitting the ground at a faster and faster pace, and the energy is the riotous energy of a bull market, so you want to be up every day before the sun is, to catch it.

During the dry spell, Mark had spent several days behind a team of two horses, plowing twenty-four acres for corn. We'd graduated by then from a walk-behind plow to a sulky plow that the driver could ride on, raising and lowering the plow bottoms with a spring-loaded lever—a leap of technology that brought us in a flash from the preindustrial world all the way up to the 1920s. The sulky plow was easier to control than the walking plow, and the two bottoms allowed us to plow more easily in both directions. After plowing came harrowing, which smoothed the furrows and was done with a heavy disc harrow we'd pulled from our neighbor Ron's hedgerow. Then the spring-tine harrow, to make a fine seedbed, and the soil was ready to receive the corn. The size of our membership was growing fast, and we needed to produce more grain to feed the chickens and pigs; if the quality of our hay wasn't good enough, we'd need the grain to bring the dairy cows through the darkest part of winter, and the horses would eat it when they were in heavy work. It also would become the sort of belly-filling food that would anchor our meals for the next year—cornbread, spoon bread, polenta, tamales, hominy, johnnycakes, and tortillas. We would plant an acre of sweet corn, but compared to field corn, that was a novelty item, a lovely treat in late summer, not the serious stuff that would really feed us.

It's hard to overstate the importance of corn—a great concentrator of resources, almost pure carbohydrate, taking the sun's energy and converting it to a gorgeous golden grain. Corn has taken its knocks as a heavily subsidized monocrop and a cheap source of calories, but it is a plant to be revered. You can feel its power when you run through a field of it in late summer, the rows higher than your head, giant garden spiders looping their elaborate webs from stalk to stalk. The long, wide leaves are palpably alive, with their razored edges breathing in carbon dioxide and breathing out oxygen so that you feel like you are running through a chamber of alien beings who exude strength. Every few rows, there's a blighted ear, like something struck by a curse, enveloped in a gray-black fungus that is ugly but in fact deliciously edible.

After the sun retreats in the fall, the green life will flow out of the leaves and they will rattle dryly on their stalks. All their energy will have run into the kernels by then, and the ears will hang down, dense and gold, wrapped in a sheath of silk and husk. Then the sun's energy is safe, stable, mobile, storable. Sometimes the raccoons discover the corn and take their share, hauling the loot down with their clever hands, or the crows, who perch on the stalks, or some lucky deer, who nibble at it and grow fat. Sometimes it's discovered by bears, who eat corn in a pantomime of gluttony. They lumber into the best part of the field and plop themselves down, then swoop out a paw, knock down stalks, and pull them close so they don't have to move while they eat. Unlike the neat deer, they leave a total mess. Half-chewed mouthfuls of corn, a rough circle of flattened plants, and, as an insult, giant piles of corn-filled poop.

If all goes well, when the ground has begun to freeze and the rest of the land is sleeping, the combine will arrive, hired from our grain-farming neighbors, the Wrisleys. It will devour the rows six at a time, cutting the dry stalks and the leaves, grabbing the ears with its metal teeth, scraping the kernels from their sockets, leaving the cobs behind in the dirt. The kernels will become a river of golden energy that flows out of the combine into the bed of the truck that travels next to it. Then to the auger, which lifts it to the top of the grain bin in the farmyard. The corn will sit there like a dense cold star, sleeping, waiting to be ignited by teeth or beak, and be transformed again in the cauldron of an animal's stomach.

• • •

Mark and I woke up before dawn and choreographed corn-planting day together, in bed, a particular form of intimacy that we'd had since we met. The farm was so central to our marriage that I sometimes wondered what other married people talked about or did together in the hours they weren't at work. All farms are busy in late May, but on ours, with its extreme diversity of products, spring can be absurdly complicated. We had six people and eight draft horses to deploy that day. Corn planting was the clear priority. But there were always the regular chores to do, the morning and evening milking, and today, the first cultivation of the transplants, which had taken root after the rain, as had the weeds. If we didn't hit the weeds now, they'd grow too rank for the horse-drawn cultivator and need to be pulled by hand.

First, though, we needed breakfast. It was half past five, so we heaved ourselves out of our warm covers, put on our work clothes, collected our pocket tools from the tops of our dressers—a Leatherman each, our notebooks, pens, and earplugs, plus two headlamps—and walked downstairs.

Blaine was already in our kitchen, boiling water for an enormous pot of coffee. She'd come to work for us the previous summer, her long legs unfurling out of a little red Honda Civic hatchback that was even older and more beat-up than ours. She was twenty-five years old, nearly six feet tall, with perfect olive skin and thick brown hair that fell in waves to her waist. She had tribal tattoos around her forearms. She was strong, stubborn, opinionated, and hot-tempered. I'd never met a woman who cared less about what other people thought of her. She had spent the previous seasons on a farming collective in the Hudson Valley, run by artists and anarchists. I'd heard a rumor that she'd been kicked out of the collective, which, in retrospect, was a pretty big red flag. But as our farm grew, the thought of hiring someone whom anarchists couldn't live with was less intimidating to us than shouldering the increased weight of the work. Over the winter, I'd become convinced that hiring Blaine had been the right decision. She worked hard, used common sense, knew her way around tools, and was fresh out of welding school. She was most interested, though, in animals and their meat, so she'd become our butcher. I liked her, even the fierce parts. So did Jane, who would spend afternoon hours in front of the butcher shop, watching her at her bloody work. I think Jane liked Blaine because she never spoke in the childish voice that other adults use with kids; she addressed Jane seriously, as a colleague or a peer. Blaine camouflaged her looks in coveralls and thrift-store shirts and never wore makeup.

Sometimes she went long stretches without brushing her hair, until it hived up in thick clumps. But still, when new people arrived on the farm, they were almost physically arrested by the sight of her, walking down the road with a rifle over her shoulder, or at the butcher shop, peeling the skin from a gutted hog with her sharp knife.

When the coffee was rolling, Racey arrived with Tim, and they laid an enormous griddle over two burners of the stove and began frying eggs in a slick of homemade butter, two dozen at once, the shells flying through the air in the general direction of the compost bucket. Tim was all effortless health and cheekbones and beautiful hair. Between college and graduate school, he'd been a cliff diver in Hawaii. He could do a standing back tuck—a skill he deployed once from the back of a hay wagon on the county highway when the horses lurched unexpectedly forward; he landed gracefully on his feet. He was sunny, friendly, thoughtful, and kind. He had just finished a master's degree in rural anthropology, and he came to work for us because he wanted his own farm one day and thought he wanted to work with horses. Soon after he arrived, he bought himself a young green-broke team of Percheron geldings, dapple gray and not yet filled out but lithe and tall.

At the table, Chad was slicing a loaf of the sourdough bread I'd baked the day before, and slathering the slices with butter to go under the broiler for toast. He was a few years older than the others, strong and calm, laconic, but funny when he chose to tell a story. He had a quirky accent that was half central New York, half rural Virginia. He'd gone to forestry school nearby but had been working for the last few years as a horse logger in Virginia. Horse logging is a tough business even in good times. Expenses are high, and the work is hard and risky. The market is small and specialized: the small subset of people who will pay extra to log their land more slowly and thoughtfully. When the economy tanked in 2008, the market dried up, and Chad couldn't make a living with his team in the woods. So he'd come to work with us, bringing a lovely, steady team of chestnut horses named Fern and Arch, whom he'd raised from foals.

Racey, Blaine, Tim, and Chad lived together in the Yellow House, a once elegant three-story nineteenth-century home next to the village library, just across the street from the ferry dock. It had been vacant for almost a decade, after the death of its last occupant, and was due for a gut renovation, so the rent was cheap, and nobody minded the muddy boots piled on the porch or the bits of straw that clung to the armchair's yellowed lace antimacassar. They were first-generation farmers, like we were. Mark and I had some years of experience on them, but without the underpinning of tradition or the certainty of generational knowledge, we often felt like we were making it up together as we went along. We were all entirely committed that year to creating what we had imagined, this system that took in work and cranked out food.

The toast, eggs, and coffee disappeared as we sorted out the jobs. I could bring the cows in for milking with my dog, Jet, then catch the four horses Mark would need for planting corn. Mark, Racey, and Blaine would start milking. Chad would hitch his team to cultivate the transplants, and Tim would handle the animal chores. I would get Jane some breakfast, then I could start lunch for the whole team. Later, Jane and I could collect the eggs and move the fence to give the beef cattle fresh pasture. The important thing was the corn had to be planted by the end of the day. The forecast called for two inches of rain that night. Our farm's fatal flaw was poor drainage. The best soil—that beautiful dirt from the glacier's nose—sat on flat land at the bottom of a hill. Two inches would keep us out of the fields for at least a week, during which the window for corn planting would slam shut. If we were going to plant corn, it would have to be now. At the end of the meeting, we all put our fists together in the middle of the table for the daily cheer—farming is fun!—then broke to jog, like a football team, to our starting positions.

Jane was still asleep, so as the rest of the team cleaned up breakfast, Jet and I walked outside to bring in the cows. Jet was an English shepherd, a gift from Mark on our first Christmas as married people. He was large and heavy-boned for his breed, black with white details and a thick mane of fur around his neck that made him look regal and masculine. He took the duty of adoring me as seriously as any of his other work. Wherever I went, Jet was in my shadow, his long feathery tail slowly waving over his back. He'd move cows if I needed him to, catch chickens, kill woodchucks, chase off crows, or do his very best at any other job I could make him understand. If there was nothing for him to do, he was happy to lounge in the shade next to the house until something came up. He was a laid-back, intelligent, all-around working dog. In his old age, his talent for love would stop being specific to me and generalize to every person who came to the farm. His retirement job was farm diplomat, an important position to which he was perfectly suited.

The sun was just coming up, casting a soft, flattering light on the landscape. Everything we had in use that spring was somewhere between well-worn and worn out, mostly functional but held together with tarps, tape, and twine. By then I'd learned to squint to filter out the farm's flapping and rusty parts and see instead the glorious edible landscape we lived in, the mountains to our east and west, the glimpse of the lake between green leaves, the animals on productive pastures. This squinting, metaphorically speaking, is a skill as important as any other on the farm, to keep the focus on the rewards and blur the hard parts. We'd bought the land the same year Jane was born, but I still couldn't believe we had a deed that said all this was in some way *ours*. I felt about the land the same way Jet felt about me. Loyal.

South of the house, there was an open-sided pavilion—a cracked concrete pad with a roof over it that had been at the farm when we'd arrived, and where we set out each week's bounty for our members; next to it was our butcher shop, which we had made from an old tractor-trailer box because we didn't have the money to build something better when we'd needed it. We'd bought the used insulated trailer for fifteen hundred dollars, sold the wheels for five hundred, then added electricity, running water, racks on the wall for butchering knives, and a track on the ceiling to move heavy carcasses along on a hook. The outside walls were beginning to rust, but we'd gained some natural light inside after an especially claustrophobic butcher had cut a crude window to the outside world. Like most of the farm, it was not perfect, but it worked.

Jet and I turned north, past the machine shop and the pole barn, along the pitted driveway, toward the two main barns. The East Barn was the newer of the two, with low ceilings on the first story and a large loft above. Its red paint was coming off in moth-size flakes that fluttered around the perimeter. The West Barn, next door, was its older and more elegant sister. It had been built of heavy hand-hewn timbers in 1890, as a stable for the farm's draft and buggy horses. Back then our land had been owned by the son of a U.S. Supreme Court justice, a gentleman who aspired to be a farmer. He had built himself a stone mansion on the lake, made a demonstration farm of the thousand acres that stretched behind it, and stocked the pastures with pretty Guernsey cows. He'd spared no expense, even building a round barn with a slate roof to house his cows and their hay. Nobody builds a round barn because it's practical; a square angle is cheaper than a curve. But the round barn had been beautiful: I have a postcard of it in its glory days. Though it burned down in the 1940s, some of the older people in the village remembered it. And we could still see its circular ghost, stretched between the two stone ramps that were its foundation, on either side of our cluttered barnyard.

The quality and beauty of the round barn lived in the beams and sound walls and good light of the West Barn. Mark and I got married in its cathedral-like loft. Every year since then, it had held hay. As the summer ticked by, we would fill it with fragrant bales, and as the winter progressed, we would empty it again.

The dairy cows were lying down in Long Pasture, northeast of the barns, chewing their cud. They raised their heads as Jet approached, then stood and stretched, and began to plod along the path to the barn. They knew the routine. Jet was merely an incentive, weaving slowly behind them, calm as a monk.

We were milking seven Jersey cows that year, and most of them were three months bred, compliments of a beady-eyed bull named Brian who appeared regularly in my nightmares. Jersey cows are known for their soft, sweet natures, but Jersey bulls are notoriously dangerous, especially as they mature. There's a short window between when they are useful and when they are dangerous. They aren't tall or mature enough to breed cows until they're about a year old, but by eighteen months, they've become increasingly unpredictable. Brian was usually calm and compliant, but one dark and sleety evening the previous winter, I had been bringing in the cows with Jet. One of the cows was in heat, and Brian decided he didn't want me to take his girlfriend to the barn. He dropped his head, pawed the dirty snow, turned broadside to me, and bellowed. It was no empty threat; dairy bulls kill people regularly. When Brian shook his head and feinted at me, I realized I'd forgotten the stout stick I usually carried in the pasture with the bull, and I started to calculate how long it would take me to reach the fence at a dead run. Then Jet appeared out of the dark, a black-and-white streak. I'd never seen him move that fast. The easygoing dude was gone, replaced by this ferocious bruiser. He launched himself at the bull's face and bit him on the nose. Brian shook him off, turned, and trotted to the end of the field with Jet snapping at his heels. Then Jet reversed direction, came back to me, and brought the cows in quietly, as though nothing at all had happened.

We walked the cows into the West Barn through an awkward cinder-block addition from the 1980s. The roof leaked at the junction, and the windows were too small, so that section was always close and dimly lit. We'd patched the leaks on the ceiling with old vinyl billboard tarps that had proved surprisingly durable. Inside, we used more billboard tarps as giant curtains that we could raise and lower to section off the barn—milking stanchions in the south end, draft-horse stalls in the middle, farrowing pens in the north.

Each cow had her own stanchion, with her name hung on a card above it, and knew exactly where to go. Delia had the place of honor by the door, and I paused to scratch the root of her tail. She had been our first cow, back when one cow was enough, and I had learned to milk on her.

She was ten years old now, officially old for a milk cow, and her daughters and granddaughters now graced the lineup. Her fawn-and-white-spotted coat was sun-bleached to a vague tan, and her udder swung loose and droopy when empty, the teats hanging well below her hocks. She looked warily at Jet. Delia's dislike of dogs, even gentle ones, was justified. She had been attacked by three dogs after arriving at our farm. The dogs had belonged to the tenants then in the farmhouse, and they had nearly killed her. She still carried the scars—the most obvious being her lack of ears. The dogs had ripped them to shreds that our veterinarian, Dr. Goldwasser, had trimmed cleanly away, so that she was left with a set of small waxy nubs on the sides of her head. My affection for Delia was due in part to what she'd been through, in part to the good milk she had given us, and in part to the fact that she was the only creature on the farm who had lived with us through all the changes of the past seven years. However, my affection would not be enough to extend her life another season. I'd learned by then that on a farm, it was not only possible but necessary to love an animal and also kill it. When Dr. Goldwasser had come to check the dairy herd for pregnancy, the rest of the cows were bred, but Delia was open. Who knew why she hadn't conceived—age, a difficult calving, some minor nutritional stress, hard weather? Whatever the cause, she would not have a calf that year, and without a calf, there would be no more milk. Moreover, that udder: it hung so low, it was hard to keep the teat ends clean. She'd had a bout with mastitis over the winter, and it was only a matter of time before another. When her milk production slowed, toward the natural end of this lactation, she'd be culled. Cull, from the Latin *colligere*, to collect. Her final gift to us would be her muscles, organs, and bones. The decision to cull a cow is a choice against the individual and for the whole. It gives me a cold, sick feeling to make it, and after it's made, there's relief.

I checked the cow's condition as they filed into their places. Three weeks ago, they'd all seemed winter-weary, with grass-longing in their eyes and their bags less than full at milking time. Now their coats had made the transition between the thick fullness of winter and the sleek shine of summer, and they came into the barn with tight udders. Still, they could use some weight on them. Early spring grass is full of protein but not a lot of carbohydrates. The protein boosted their milk production, but the calories required were more than this grass provided, so they pulled what they needed from their own body fat. Their hips had begun to look sharp. Some sweet, succulent roots would do them good.

I could hear Mark in the milk house, gathering the pails and cans. Our first year, when we had cows to milk and members to feed but no infrastructure, we'd milked into buckets, filtered the milk into ten-gallon cans, and chilled them in ice water. There was something very satisfying about that short straight line between cow and consumer, which required only about two hundred dollars' worth of equipment and provided dairy every week for thirty people. We washed the pails and the ten-gallon cans in the sink in our house. One day the milk inspector paid us a surprise visit and pointed out the ways in which this was illegal. It was fine to milk by hand if we could do so cleanly, and as long as the milk got cold enough fast enough, the ice-water chilling was okay too. But we were supposed to have monthly inspections and analysis of the milk's bacteria count, screenings for pathogens, plus a license to supply raw milk. A license required a proper milk house adjacent to the barn. We didn't have nearly enough money to build something new, so we called the same man who had sold us the tractor trailer that became the butcher shop, and he brought us another one, drove it up to the wall of the West Barn and dropped it there. We sold the wheels from that one too, cut a hole in the back gate for a door, plumbed and wired the box for hot water and electricity, and filled it with sinks and racks and the stainless-steel pails, and boom, we had a milk house and were back in business within a week.

That morning Mark had shouldered fifty-pound bags of carrots, beets, and turnips to the root grinder. The root harvest had been excellent the previous fall, and we were feeding the cows what our members could not eat. This was part of the beauty of producing a full diet year-round. Waste from one part of the system became fuel for another. The grinder was an antique thing we'd found at the back of the barn when we'd moved in, a relic from a time when this kind of synergy was normal. It was in surprisingly good shape, its bright red paint mostly intact. It had a stout handle attached to an axle that spun a cylinder inside, which was notched all over with sharp moon-shaped blades that bit into the roots, cut them to chips, and spat out a giant's chopped salad. We had dug the roots in late fall and stored them deep in the buried stone foundation of the old round barn. The bags at the outside edge had frozen on the coldest nights and stayed that way all winter, so in the spring, the orange chips of carrot were slightly mushed on the edges and crystal inside. Mark could make cutting through those hard roots look easy, his long strong arms whipping the cylinder around at high speed, but I knew the weight of the handle. In the winter, no matter how cold it was outside, the work of turning it against its heavy gear would heat me from the inside. If I lost control of the handle, it could knock me across the aisle of the barn.

The cows were eyeing the roots impatiently, so I stopped to serve them by the shovelful in front of their stanchions. As they crunched the roots, the beets stained their lips red, and they looked like they were wearing sloppy lipstick, a line of tipsy matrons out for a night on the town. Racey sat on a stool and got to work. We milked by hand then. Our mornings began in the dark, faces pressed into the fragrant flank of a cow, warm teats in our hands. It was a soft time of day, for quiet talk. When there had been only four cows, Mark and I milked alone, two cows each, and the barn at dawn felt like an extension of our bedroom. We told each other our dreams, and they felt close and real in the dim light. Or we talked out our plans, small and grand, for what the farm could be. As the membership grew, we added more cows and needed more people to milk them. The loss of intimacy between us was balanced by the relief from overwork, not just in the dairy but on the whole farm. We had employees by then, but the growth curve of work was always slightly ahead of the available labor. Sometimes our friend Jay, who taught science at the high school, would run the five miles between his house and our farm to milk two or three cows before his first class. In the evening, his wife, Kristin, would come on her own or with her youngest child, who was two years older than Jane.

Our membership was growing again that spring, and we needed more cows. We were going cow shopping that weekend to look at a lot of eight Jersey heifers, all about to calve. That would more than double our herd and our production. It was time to switch from hand-milking to machines. Sweet as that work was, I would not be sad to see the end of it. My arms and wrists had had enough. Some nights my hands went numb or throbbed enough to wake me out of a deep sleep. Mark had spent the previous week in the barn, installing the pump for the two bucket milkers we'd bought used for six hundred dollars. For years I would call it the best money we'd ever spent.

As the others started milking, I dashed back to the house to check on Jane, who was still sound asleep, then looped five halters and lead ropes over my arm and walked north again. The nine horses were nipping green shoots in a paddock close against the edge of the sugarbush, which rose on a hill to the west of the house. Two were Chad's and two were Tim's; the other four and the pony were ours.

Jay and Jack were our experienced team of horses. They were in their mid-teens and still worked regularly but were getting too old for long hard days and the growing amount of land we had under cultivation, so we added a young team of Belgians, Jake and Abby. They'd arrived the year before, broke to work but without much polish, and had spent the previous summer learning their jobs, which ranged from plowing in the spring to hauling firewood in the winter. They were in their prime, lean but strong, with shiny coats and bright eyes. They looked like the pictures I'd seen of farm teams from the 1920s and 1930s, after farmers had scaled up the size of their equipment and their horses to match, just before the tractor arrived. When the tractor came, draft horses were no longer central to the work of most farms, and the horses who remained got huge, with fat necks and humongous feet, as though we wanted a sense of massive power, something to match the diesel that propelled the tractor across the field. The look of draft horses went from sturdy with moderate frames to tall, heavy, and high-headed. Of the breeds that originated for farmwork, only the Suffolk remained true to its original type. Suffolks were compact and muscular, with famously good hooves and sensible temperaments. They were rare and expensive. Chad's team was full Suffolk, and Jay and Jack were a quarter, crossed with Belgian.

I buckled the halters over their heads and knotted the lead ropes loosely around their necks while I caught Jake and Abby. Mark would need the older team's experience and all of the younger team's strength to get the corn planted by the end of the day. I unknotted the four lead lines from around the horses' necks and strung them out, abreast, two on one side of me, two on the other, with the pony like a nippy little sidecar at the end. We walked to the barn that way, and I tied them into their stalls, which were freshly bedded with straw, a pile of hay and a water bucket in front of each one. I brushed them quickly, focusing on the shoulders, where the collar sat, and the parts of their backs that carried the weight of the tongue. I lugged the four collars and the heavy harnesses one at a time from the tack room, standing on my toes to shove and nudge them onto the horses' backs, which were higher than my head. I left the britchen loose over their backs so they could stretch forward to nibble their hay. Mark could bridle and hitch them when he was ready to go to the field. Then I hurried out of the barn to make breakfast for Jane.

Every food has its moment of perfection on our farm, and every moment has its perfect food. Asparagus cedes the podium to astounding spring butter, then the first radishes, the first strawberries, then new potatoes, and so on in a harmonious stream. Our grandparents and great-grandparents understood this, of course; it's just the advent of the global food system over the last eighty years or so that dulled our sense of which food comes in which season. If you had asked me, before farming, when pears were fresh in the Northeast or when the shell peas came, I would not have been able to tell you. They were always available, disconnected from any season or geography. Now those foods are bound to my sense of the year as tightly as Christmas is to December, and the thought of eating them out of season feels as weird as decorating a Christmas tree might in July. But there are times of the year that offer abundant choices and times that are leaner. The bottom of the year comes in spring, between the return of light and the arrival of heat. That's when the wild plants have their moment of perfection.

A patch of stinging nettles grew in the lee of the West Barn, in pockets of rich soil, where one of the farmers who had worked this land before us had piled the winter barn scrapings from his cows. This was the warmest place on the farm in early spring, sheltered from the wind. There were other early, wild greens coming up. The first garlic mustard was growing in the remnants of last year's kitchen garden, around the patch of scrub left behind after Mark cut down the black walnut trees, just south of the farmhouse. Lamb's-quarters were sprouting across the driveway, where the pigs had been pastured last summer. There would be tiny shoots of purslane emerging between the rows of last year's tomatoes, but it would be weeks before they matured into something succulent and edible, and by then the civilized plants would be growing.

I took off my jacket and cut the nettles into it, along with some garlic mustard, and walked back toward the house, stopping at the nest box in the East Barn on my way. The hens had been awake for hours, foraging and scratching in the worm-rich pasture outside the barn, and were becoming preoccupied with the serious business of laying. Eggs are a special kind of magic. Forged from grass, worms, insects, and grain in the mysterious depths of the hen, they appeared like gems in the nest box each day, cased in flawless shells that were both fragile and strong. You have to put a hen-warm egg against your lips to fully appreciate its particular texture. Inside the perfect packaging lurks its slightly creepy embryonic truth. It's an animal nut, not life yet, but the rich seed of life. It holds the instructions for feather and nail, beak and brain, scratch and cluck, lacking only a little more magic to make it so—the heat of maternal love. I dug under a hen and found a whole clutch, pocketed six, and threw a cracked one to the pigs, who had wintered in the run-in on the east side of the barn. There were five sows with a litter of piglets each, nested in deep hay. They were thin with the work of producing rich milk for those demanding babies, and they squabbled over the treat.

In the kitchen, I heated some lard in a cast-iron skillet, chopped an onion from the root cellar, and added it with salt and pepper to the sizzling pan. I put on rubber gloves to pull the nettle leaves from their stems and chop them. Raw, they packed a good sting, but as soon as they hit the heat, they would wilt and lose their fierceness. I added them to the pan along with the garlic mustard. The nettles sent up their green and slightly nutty scent, like spinach with a big personality.

I whisked the eggs with a huge heaping spoonful of thick cream from the dairy, more salt, pepper, a pinch of fresh nutmeg, and poured them into the pan, then sprinkled the top of the eggs with some grated cheese cut from a five-pound wheel in the cellar. I put it under the broiler just as Jane was waking up. When she came thumping downstairs at seven o'clock, the eggs were puffed and brown on top, and we ate the farmer soufflé together, the taste of spring and of home.

Mark didn't start planting the corn until well after lunch. A dozen small things had gone wrong and delayed him. The hydraulics on the forecart weren't working, and then he discovered one of the coulters on the planter was jammed, blocking the flow of seed. As soon as those things were fixed, the sows got out, and I needed him to help me and Jet get them back in their pen before they destroyed the soft spring soil in the next field. Then it was lunchtime, and the whole crew came in to eat the hasty lunch I'd made—chili, with a big skillet of cornbread and a chopped cabbage salad, old standbys I could make fast with what I had in the pantry and freezer. At the end of the meal, when the dishes were cleared and the black coffee drunk and the pots hung back up over the barbarous sink, the floor of the dining room was littered with snoring people taking a ten-minute snooze before heading back to the field. The horses waited in the shady barn, harnessed, pulling at mouthfuls of hay. The clouds were rolling in, and the wind was coming up. Finally, everything was ready. I helped Mark lead the horses out and hitch them, two on either side of the forecart's tongue. To steer four horses with only two lines, you give up some precision for simplicity. The outside horses had both sides of their bit attached to the lines, but the inside horses had only one side each; the other sides were attached to the hame of the horse next to them. Driving larger hitches of horses felt like driving a barge or a semi. Any change in direction happened slowly, so you had to think ahead. The good part was that the horses tended to be calm and tractable in larger hitches. They are gregarious animals, after all, and four of them hitched close together is a sort of artificial herd.

While Mark loaded the seed corn onto a wagon, Jane and I went to the barn for Belle and two five-gallon buckets. This was Belle's side gig:

when she wasn't the child carrier, she was the egg pony. The pullets were pastured half a mile from the barn and were just beginning to lay. I had fashioned a complex carrying system for Belle out of a bareback pad, two buckets, and a bungee cord; it meant making sure exactly half the eggs were in each bucket, so things would stay balanced. A hundred eggs in a rickety system on a spooky pony was not the best idea I'd ever had, but I was trying to justify Belle's existence on the farm to Mark through full productive employment. In the beginning, she was skeptical about the flapping birds, the electric fence, and the buckets banging around her sides, but I had sweetened the deal by giving her a handful of chicken feed every time she was tied up among the birds, while I collected the eggs. Pretty soon she not only stood still for egg collection but looked forward to it. Ponies are essentially ruled by their stomach.

Jane and I hitched a ride down to the layer flock on the back of Mark's wagon, Belle trotting along behind. The first fit of rain had passed overhead, dropping a quarter of an inch and turning the warm air to a thick haze, the ride down the hill was a pleasure, with Jane nestled into my lap. The pastures were crazy with birds. We saw a pair of bluebirds, and then the black-and-white flash of a bobolink, who was filling the air with his loopy metallic call.

As we passed the beef herd, I spotted a problem. Through the hedgerow, I could see a large puddle in the middle of the field and a fountain of spray coming from a leaking water line. A few cows and calves were standing downwind of it, enjoying a refreshing spritz. Shouting to Mark to whoa the horses, I jumped off the wagon, lifted Jane down, tied the pony to a fence post, and went to investigate while Mark went on to seed the field. There was a small hole in the hose, a straightforward repair that would require some tools from the machine shop. But when I went back to get the pony, I found an empty fence post. I had the same feeling I used to have in the city when my car got towed: *Maybe I forgot where I parked. It has to be around here somewhere.* Then the slow dawning that indeed it was gone.

Jane and I trudged back up the hill on foot. Racey, Tim, and Blaine were in the barnyard, wrapping things up for the day. They hadn't seen a stray pony, but they were willing to stay late and help. Between us, we fixed the hose, collected the eggs, took Mark his dinner in the field, fed the hungry toddler, and finally found the errant pony, who was standing innocently in a stall in the barn. At dark, Tim drove a tractor down to the field to take the place of the horses, who had no headlights, and took the weary horses home to unharness for a good brushing and some feed.

Mark made it back to the house just before midnight. I woke up to find him sitting on the edge of the bed, headlamp still on, smiling. His face and clothes were streaked with mud from the combination of field dust and the rain, which had just arrived in earnest, pounding on the farmhouse roof and watering in the seed.

There'd been a steady stream of frustrations that day but also a lot of luck. The corn planter's hitch had broken clean off, Mark said—the sort of definitive break that forced you to stop work and would take hours of welding in the machine shop to fix. But it had happened just as he was rolling out, two feet *after* he'd finished seeding the field. The rain had started then too, and the timing of those things delighted him. He laughed, telling me about it, through the mud and his own exhaustion. That was the way Mark always told the story of his day: with the focus not on the failure of the hitch or the mud on his clothes but on the great luck of the timing. It felt like a metaphor for the whole farm that year. We faced a series of problems and frustrations, but they were lucky problems. We'd created a life together that presented to us precisely the problems we wanted to spend our time solving, and we felt up to the challenge.

CHAPTER 4

The corn germinated and grew. The greenhouse filled up with new seedlings and then emptied again as the summer crops went to the field. There were savoy and Chinese cabbages, broccoli, kale, and lettuces of all kinds. Next to the house, we'd planted an acre of buckwheat to control weeds and build the soil. Every morning, as soon as the sun rose, I'd walk out with my coffee to see how much it had grown overnight. Buckwheat is a sprinter, leaping out of the soil fast enough to outpace all the other plants, its succulent leaves vigorous enough to shade weeds into stunted things that will not go to seed. One morning, instead of green, it was white with blossoms and already loud with bees.

July came. Haymaking time. Long days under a scorching sun. The first cherry tomatoes ripened. I went out barefoot to pick them in the afternoon, and the soil was so hot it burned my feet. The corn came on strong and tall, soaking in all the heat, turning it to leaf and stem. We started those days at three in the morning. We hitched the horses in the dark and came out of the barn at four under the first gray hint of dawn. Even then the horses were damp with sweat. We quit at lunchtime, slept in the shade in the afternoon, then picked up again in the evening. By the end of the week, everyone looked wrung out. Everyone except Mark. Like the corn, he seemed to thrive in the heat.

The tomatoes rolled in by the hundredweight, the green beans stacked in bins the size of ottomans. The rain came exactly when we needed it and cleared away when we didn't. The pastures grew, were grazed, and regrew into lush carpets of grass and clover, to be grazed again. My belly grew too, every day closer to ripe and every day a little heavier to carry. It was uncomfortable, but by then I knew uncomfortable is not the worst thing in the world. Sometimes you have to do things when you don't feel strong or well. A farm, with its shifting chorus of constant needs, won't let you slide. We forget sometimes that uncomfortable is not the same as harmful and is something humans can get used to. I was pretty certain that for me, it was better to have life animated by clear purpose, and structured by a farm's responsibilities, than to always be comfortable.

The people who came to work with us seemed to feel the same way. It mattered that the work was physical and that it happened outside. For all of us who lasted more than a week on the farm, working outdoors made us happy. Working with our bodies made us happy. I thought of Jane at two in a temper tantrum, gone boneless. That state came from an overload of frustration and was temporary. Nobody really wants to be boneless. We have bodies, and it generally feels good to use them. It feels especially good when we use them for a purpose, toward something we believe in. It is a simple thing but profound, and it is easy to lose sight of in a world that places more value on work that happens inside, on a screen, while seated at a desk. "If people knew how much fun farming is," Mark used to say, "we wouldn't be able to keep them away."

Not everyone is susceptible, though. Plenty of people who came to work on our farm found it had no magic to offer them. To some people, farming was simply ill-compensated drudgery or struggle and decay. If you didn't see the work, at least in part, as something more than a job—as a calling—it made no sense to keep going. We didn't pay well, and people who weren't used to sweat and sore muscles would quickly be miserable. One woman came all the way from California, worked a day, and drove off in the middle of the night, leaving a scribbled note of explanation on our door.

There are weeks in May when anyone in the world can find pleasure in farmwork. There is bright sun, the smell of warm earth, a cool breeze. The horses trot out of the barn eager, fueled by the sweet young shoots of grass, glowing from the inside. It's all so lovely and wonderful. Then, as summer comes on, things shift. The sun gets serious and the weeds begin to grow fast, threatening to overtake and smother all the work of winter and spring. Our skin burns, and we sweat, and the sweat stings where the skin is burned. You get used to it and find the rewards override the discomfort, or you get out.

Occasionally, the discomfort really is too much, for us or for the animals. The week the corn tasseled, the flies descended. We'd had a stretch of warm weather followed by a soaking rain that gave me a special feeling of dread, because that pattern is the signal for the green-bodied horseflies to hatch. They are the size of fat raisins, with helmeted eyes, like fighter pilots from some evil future realm. They have their own focused plan of work and are just doing their jobs like the rest of us—they need some blood for their babies.

I was too pregnant by then for cultivating. Sitting on the cultivator, behind two horses, I couldn't see past my belly to the rows passing between my feet. But I could stand on the forecart and drive a team to ted the hay. Tedding was a relaxing job, a light pull for the horses, that didn't require much precision. We ted after we mow, to fluff the grass and help it dry so it can be baled before the next rain comes. It has to happen quickly, in the window between rains.

I chose Jake and Abby, the young team, harnessed and hitched them, and headed up the sugarbush hill toward the field on Middle Road. It had rained the night before, and the ground was spongy. Hot, damp air was rising from the tall grass. When we brushed through it, the biting flies rose too; the pair of warm-bodied horses, already damp with sweat and exuding great clouds of attractive carbon dioxide, were the answer to their most ambitious dreams. The organic fly spray we used on the horses didn't do much. The fly nets they wore—which went over the tops of their bridles, with black strings that dangled over their eyes so they looked like they needed their bangs cut—helped a little, but only at first. The newly hatched flies were too ravenous. They bit the horses on their flanks, their necks, on the gelding's sheath and the tender place at the base of the mare's udder. The horses, in their harnesses and tightly hooked to my cart, had little recourse. They could swish their tails, shake their manes, stomp. But mostly, when working, they just had to tolerate it.

Abby became more and more frustrated. She pinned her ears. She trotted a few paces, but I used my hands on her bit to keep her from going faster. Her energy built and had to go somewhere. She dropped her head as far as the check rein would allow, kicked out with her back leg. "Quit it," I said, even though I sympathized. "Come up." We continued up the hill. Another swarm, more bites, and enough was enough. Down went Abby's head, out flashed her hoof. And this time it landed on the wrong side of the tongue of the forecart—the stout ash beam that rested in the eye on the yoke between their shoulders—so that she was straddling it. *Not good,* I thought. But after seven years, hundreds of hours behind horses, and a few terrifying incidents, I had developed some tools for situations like these. The first is to shove aside any inkling of panic, to convey only certainty and control. *Everyone just calm down,* I said in my head. I told them to whoa, set the brake on the forecart, and hoisted myself to the ground, arms crossed over my big belly.

Abby was bucking now, small, tentative bucks,

weight rocking from her front end to her back, accompanied by a high whinny like a muffled scream. The tongue was scraping against her inner thigh as she crashed up and down. No horse likes to have things between her back legs. For some, it's intolerable. Abby, I knew, could handle it, but I didn't know how long, especially with the flies taking advantage of our stillness. Meanwhile, Jake was starting to dance a little bit, upset by the way Abby was forcing him into the outside tug. He blamed her, and laid his ears back.

Everyone just calm down, I thought again. *We can work through this*. I had the lines in my hands and shortened the right one just a bit, so if they bolted, their heads would be going that way, and they'd run off the trail into the woods instead of straight up the path and onto the road. All those motions were automatic by then. So was the voice projecting calm.

The only way out was through. I had to unhitch them from the forecart so I could get Abby's leg back over the tongue. With all that explosive energy building up between them, I had to do it carefully, like a sapper clearing a mine. It's dangerous to put your body in the zone between the horse and whatever they are pulling. If they bolt, you get flattened. Very simple. You can't always stay out of that danger zone, but you must always be aware when you are in it. My belly put me at a disadvantage. I wasn't nimble.

I used my knee to push the evener forward and give myself slack to unhook Abby's outside tug chain. The chain came loose, and the evener swung back the other way. I reached around her rear end and unhooked the inside tug chain. She stood. Good mare. Now she would have more freedom to move forward, to swing her butt back into position, but if she ran taking Jake with her, the tongue would knock loose of the neck yoke in front, maybe break its safety chain, drop to the ground, dig in, disaster.

We're fine. Everyone calm down. It's a form of telecommunication. It is faking it. But it has real utility, with animals, with children, maybe even with spouses. Think it and make it so. It helps to know your horses and have their trust, which is hard earned and easily spent. This time, it worked. I unchained the tugs on Jake's side, and now I could reach Abby's leg. My head was more or less even with her back when I stood up. Her hind leg weighed nearly as much as my whole body. "Lift," I commanded, and she did, her heavy hoof dwarfing my palm. She had to flex her leg backward, hard, from the hip and the hock to get her hoof over the tongue, an uncomfortable, unnatural position. We worked it together. The toe of her hoof scraped white paint from the tongue. Then it was over. Relief. We were all in our places. I rehooked the chains, took off the brake, and we were a unit again, coordinated, cooperative, under control. The horses were as relieved as I was. They were law-and-order creatures in the end, happy to be back inside clear boundaries. Before I got onto my seat again, I walked a wide circle around them, crushing the flies on their bellies, feeling the crisp bloody pop between my hand and the horses' bodies, which were slick with sweat. Then we walked the rest of the way up the hill to ted the field.

A thing about farming that is difficult to explain is that it requires you to wield authority. The working relationship we have with animals is different from the pet relationship. Once I was visiting a man in Paris. I was in my late twenties, and I had flown there to spend a week with him, in his garret looking out toward Notre Dame. There were thick white sheets, small French appliances, the smell of day-old coffee grounds in a press. It was November. The sky was gray all week. Cold wind blew against mullioned windows. Those years were confusing ones. I was unsure of the form my life was supposed to take, unsure of pretty much everything. He was a poet. Maybe a poet could pull me through. He was only slightly older than I but infinitely surer. From a distance, over the phone, that had seemed so attractive. But in his flat, up close, I was intimidated—by his erudition, his awards, the book with his name on the cover. He did nothing to reassure me, which was not as unkind as it felt back then.

While he wrote in the mornings, I looked at paintings in the Louvre. I walked along the Seine in my high-heeled boots and black coat, which was too thin for the weather but the most stylish one I owned. When he was finished writing, I could go back to the apartment.

As the week wore on, I could feel his strong brain searching mine for something of value. That shut me down, made it difficult to speak. In bed one night, he asked me to tell him about something I knew that he didn't. He might have been trying to put me at ease, but it felt like a test. I had nothing to offer. Music, art, books. He knew more about all of them. *Horses!* I thought. Suddenly, I could talk. I spoke the parts of a horse's body, the names of the tack we use to communicate with them; I explained how different types of bits work, some on a horse's lips, some on the nose, some on the bars of a horse's mouth. Bars, you know, gums, the place where he has no teeth. I explained the idea of pressure and release, how horses will move away from pressure—from the leg, or the heel, or the bit—and are rewarded by the release of that pressure when they do what is asked. I described how good it feels the moment a horse collects underneath you, because you've asked for a paradox, to move forward with your legs and hold back with your hands, and he understands and does it, two opposite things at once. The corners of the poet's mouth turned down. I could see my monologue wasn't going over well. He remained unimpressed. "It sounds," he said, "coercive. How is that not coercive?" That I could not say. I closed my eyes, fell silent.

Is it coercive? It is, I know, hierarchical. Otherwise, it doesn't work, becomes dangerous for both sides. It's a partnership, but there's a boss. The human must hold authority, must claim the right to rule. Benevolently, fairly, but definitely. I wish now that I had lectured the poet about the difference between authority and coercion. The right to touch the velvet-nosed beast, and to look him in the eye, is something that has to be earned and, once earned, held. Horses are like fire. Color, form, movement. And like a fire, under some conditions, they are only tenuously controlled. What would we have built in this world without horses or flame? The control of them is part of what made us human.

When the whole field was tedded, I brought the tired team back to the barn, unharnessed them, brushed the dried sweat from their hair. I fed and watered them and felt their physical satisfaction, a mirror of my own. We shared the pleasant end-of-day feeling of a job well done. That's not to say they were always thrilled to put on their harnesses and head out the barn door. Or that they would not have quit sooner if they had the choice. But farming with horses taught me that when challenging things are required, we become more complex beings. When the tasks required of them are fair and possible, and when there is also sufficient rest, food, water, and mutual understanding, then the complexity is good. When the effort is toward a shared purpose, that is the best of all.

The next day, it was time to rake and bale. It was also the Fourth of July, and the volunteer fire department had organized a celebration. When Mark and I first arrived in Essex, I was shocked to learn that when something really dire happens—like your house is on fire or you are having a heart attack—help comes in the form of a neighbor. You need a massage or your teeth cleaned, you go to a professional. You need to be extracted, bleeding, from your crumpled car lodged in a snowbank, you get a volunteer. The anonymous "they" that I counted on in the city to take care of emergencies—the messy splatter of accident, of disaster, of misfortune both public and private—had become a small and knowable "we."

I had joined the volunteer fire department the year after we'd arrived. Dave Lansing, the fire chief, had driven up in his white Chevy Silverado with the array of lights on the top, as I was planting garlic in the field next to the driveway, early November. It was raining gently, I was cold, lunch was overdue, and my legs were chafing inside my Helly Hansens. When he rolled down the window and asked if I wanted to volunteer for the fire department, I said yes without thinking. It was probably because my blood sugar was low and a simple yes seemed easier than explaining a long-winded no. Also, I had no idea back then what it would entail. Mark, thinking more deeply than I, demurred.

I went to my first meeting at the firehouse, which sat at the far end of our easternmost field. I was feeling exquisitely self-conscious. I'd envisioned a roomful of beefy men, secret handshakes, maybe some cigars. When I opened the door, I thought I'd gotten the wrong night. There were three men sitting in a line of chairs, looking at an empty table. I sat in the only open chair, which was facing them, across the table. "Not there," one of the men said. "That's the chief's chair." Blushing, I moved another chair next to the three of them, making a fourth in the row. Dave was just Dave when we met him in the village, but in this building, we called him Chief. He came from the radio room and took his place, then all the chairs scraped back, and the meeting began with a recitation of the Pledge of Allegiance. I felt like an impostor with my hand over my heart—an act I had not performed since grade school assemblies—and then we all bowed our heads for a moment of silence in honor of fallen firefighters. After the meeting, they took me to the bay and showed me the truck and the tanker, and the new mini-pumper, pride of the department, all bright red and decorated with a picture of the town's antique horse-drawn pumper, which was stowed in the last bay. The other trucks were painted with the names of former chiefs. At the edge of the bay were red lockers full of bunker gear—heavy fireproof jackets and bibs, helmets, boots, face masks, gloves. I'd been worried about being accepted, and they'd warmly welcomed me. I understand now that this wasn't exactly personal. They would have been excited to have anyone to fill out their thin and aging ranks.

What the job entailed, it turned out, was quite a bit more than I'd expected. Training was held at another department, a forty-five-minute drive through the mountains. All the trainees from the north country would be there. We would meet three nights per week for twelve weeks. I did not have time for this. The farm was too much already. But I'd promised, and besides, they'd already spent the taxpayers' hard-earned money on a new set of bunker gear specially sized for me, so I went. The time would have to come out of my sleep budget. At the last minute, Mark, perhaps not wanting to be alone at home all those nights, decided to join too.

The class was full, every seat taken. The youngest were just sixteen, second- and third-generation members. They'd looked forward to this rite of passage all their lives. They talked movingly about what it meant to be able to give back to their communities. There were older people too, who had retired and finally had time to do this.

Over the course of the winter, I learned how to take apart and put together, with my eyes closed, the air tanks we wore on our backs. I learned about rollovers and flash points, how to properly climb up a ladder, and the cool trick of how to climb down one while carrying a full-grown adult. I learned how and why to vent a roof with an ax. I overcame, barely, my intense claustrophobia to crawl blindly through a training culvert filled with hanging loops of wire that caught on the bulky gear, learning how to carefully, patiently untangle myself before running out of air. At the end of the course, the chief of the host department shook my hand and told me he'd be happy to have me on his fire scene any day. I felt a tingle of real pride.

My first fire was a barn fire in the next valley, just after my training was complete. The call came at midnight, jolting us out of deep sleep. It was early spring. No moon. Jumping into my gear, I felt a confusing mix of fear and excitement.

Mark and I could see it from a mile away, the light of the fire and the light of the trucks that had already arrived—the first an ominous yellow-and-black feral glow and the second the pure blue-white focused light of civilization. It was Reggie Carver's barn. His dairy sat along one of the dark roads near the river. Every department in the region was called out, and it was crowded with trucks and people.

We turned off the road and drove along the rough cow track toward the fire. In the darkness at the edge of the light, I could see the Holsteins, loose and milling, confused. Cows are creatures of habit. They called to each other in the dark, low mournful sounds. When we got closer, there were the sounds of the fire itself and shouting men. At big fires, with volunteers from different departments, it is almost impossible to hold a proper chain of command. There is no way to communicate. The fire is loud, the trucks are loud, there is smoke, and many, many people are running in all directions, most with their own idea of the best thing to do. Our neighbor Ron set up the pumper while Mark and I got our masks and air packs on. I don't remember who told us to go in, but we did, hauling one heavy, charged hose between us. A barn cat, skinny, mangy, with the large-headed look of the closely bred, slunk around the hot foundation, illuminated by fire.

We walked like aliens into the burning milk house, spreading water in front of us. There was the bulk tank, intact, and next to it, smoldering bags of chemicals. There were the calves chained in their stanchions, the smell of burned hair, and the sound of our friend Bradley shooting the calves who had been burned but were still alive.

Once one of the bigger departments arrived with the aerial, there was not much more for us to do. They sprayed from high above, knocking down what was left of the fire. We stood by, putting out little fires around the edges when they popped up. We watched dawn reveal the smoldering remains. The fire had eaten almost everything. In the first good light, the cows were herded across the road to Lee Garvey's barn, where they were milked for several days in rotation with his own, and then they were sold. Reggie was dying of cancer, and he would not rebuild.

We all had to play a part to make our town run. Some served on the town board, volunteered to bake for the Grange fundraisers, or worked as town justice. I trained for hazardous substances, took a class on how to set up a temporary detox station, learned the symptoms of nerve gas poisoning—this was post-9/11 America, after all. Once a train derailed at the crossing a quarter-mile from our farm. Word came across the pager that a tanker had overturned and was spewing its contents on the tracks. We geared up, ready for hazardous substances. It was, it turned out, a carload of squid moving from Montreal to New York. We briefly considered going into the salvage business, offering to compost that tank of stinking nitrogen-rich stuff, then thought better of it.

After Jane was born, I stopped volunteering as a firefighter. We couldn't leave her alone to respond to calls together, and Mark's physical strength made him more valuable on a fire scene. But I became a fire commissioner, one of a board of five elected officials who allocate the budget. Sometimes I go to the firehouse when the farm is extremely busy and I need a break from the chaos. I wrote my first book in its broom closet, moving the mop bucket out each morning and a small desk in.

The fire department's big public moment was the Fourth of July celebration. There were freshly washed fire trucks from all the neighboring volunteer departments. A good showing of antique tractors. Some local businesses made hasty floats. Bagpipers played in a ragged line. There was a delegation of veterans marching, and the men from the Masonic Lodge. Some women and kids on horseback, in Western gear. The grand marshal, our town's high annual honor, was Grace McCloud, one of our oldest residents, who rode in a classic car, waving shyly. The Methodist Church sold strawberry shortcake at the corner. There were games for children down by the lake. The fire department sold grilled hamburgers and hot dogs. Afterward, there was a spelling bee in Town Hall. But the parade was the big draw. People lined up three or four deep from the library to Bull Run. They set up chairs early to get a good view and to catch the rain of candy that marchers chucked into the crowd. Everyone turned out for the parade, from the fine houses on Main Street and the trailers back in the woods.

Mark had been asked to make an appearance that year because he had some special skills. Once at a party, just after we'd first met, I heard someone ask him what he learned in college. I already knew that Mark was skeptical about the value of higher education on a cost/benefit level, so I was interested in what he would say. "I learned to juggle," he said. The other person assumed, as I did then, that he meant to juggle *priorities,* but he'd meant it literally. In between classes, on the weekends, he'd worked on his circus tricks. Juggling was first. He had started with bean bags and moved to balls, then clubs, then knives and flaming torches. He'd learned to walk a slack line and ride a unicycle. In his intense, obsessive way, he'd zeroed in on these things, taught himself to do them really well. I was glad that I hadn't known this about him before we'd gotten married, because I was pretty sure I wouldn't have been attracted to him during the circus phase. But his tricks came in handy on the Fourth of July.

When it was time for the parade, I met him in the field, where he was raking the hay I'd tedded the day before. Someone had to keep raking, even if it was the Fourth of July. I climbed onto the forecart and took the lines, and he climbed off, retrieving his costume from the hedgerow. He'd come up with an outfit for the parade that we were calling Sketchy the Clown: a beat-up old spandex singlet that left nothing to the imagination, a pink-and-green-striped shirt, dirty white leather sneakers, and his usual everyday big straw hat. He was sweaty and red-faced from the heat and the work, and his hair was stuck with chaff. He had three juggling clubs and his unicycle. He smiled and waved at me, walking off to the start of the parade, looking less like a clown and more like a maniac.

When he rejoined me in the field a few hours later, I asked him how it went. He's usually hyperbolically positive when reporting on any sort of performance. This time, though, he looked sheepish. "Not so well," he said. He had attempted an ambitious trick and lost control of the unicycle, and one of his clubs had flown into the crowd, where it bonked a small child on the head. "Was she hurt?" I asked. "She cried a lot," he said. Then, after the family tried to move out of the line of fire, he ended up unicycling past them again, and on the second pass, he heard the father say, "Oh no, here he comes!" before scooping the girl up in his arms. Given Mark's appearance, the poor child was probably at least as scared as she was hurt. To his credit, he'd made the time to do it, even though we were in the middle of a haying marathon. We brought in twenty-five hundred bales that holiday weekend. The West Barn hayloft was mostly full, and we were at least a third of the way to having what we needed for the year.

In August, one of the beef cows gave birth to a late calf in the pasture, a little brown heifer. The beef herd was made up of cows descended from the Scottish Highland cattle we'd bought for a bargain when we were first starting out. They were tough as nails and good strong mothers. In the winter, their heavy coats kept out the cold and the snow. In the summer, though, those coats could be a liability. When the calf was a week old, Racey came to the house to say she didn't seem to be nursing. *Ugh, fly-strike,* I thought. Bane of summer-born babies, especially those with long hair. The blowflies lay eggs in the birth fluids, and they hatch into maggots that can eat a calf alive. We walked back to the pasture together with a bucket of soapy water to clean her. The cure for fly-strike is mechanical removal of all the little offenders. Catching the calf was easy. She was not doing well and couldn't run very fast. At first I couldn't find what was wrong with her. But when I lifted a mangy patch of fur on her back, I saw the skin underneath alive with masses of white wriggling worms. This was not a job we could do on pasture, so we dumped the inadequate bucket of water on the ground and hefted her back to the barn on our shoulders, taking turns, the maggots falling into our hair like living dandruff.

Dr. Goldwasser happened to be on the farm, treating one of Tim's horses for a scratched eye. He helped us shave the worst parts—a bloody raw patch the size of a football on the calf's back and another on her hind leg. It would be impossible to get all of the maggots in one go, so she would need to stay at the barn for a few days of assiduous scrubbing and to regain some strength. I went back to the pasture to herd her mother up so the calf could nurse, while Mark worked on maggot eradication, using a hose and a brush. Each time we thought we were finished, we would comb through the fur with our fingers, and one would wriggle and give itself away. It took a while, but at last it was done.

We went back to the house for long, hot, soapy showers and dinner with Jane, in that order.

That night in bed, as we were replaying the day, Mark leaned over to kiss me. He paused and scratched at his midsection, one eyebrow raised, then shrugged. Exhausted, we clicked off the light. An hour later, he clicked it back on again and sat quickly upright. “I need you to look at something,” he said, pointing. I got the flashlight and illuminated his belly button. Deep in there, something moved. It was a slender, limbless spelunker, its tiny jaws looking for nourishment. I got the tweezers and plucked it out, extinguished its soft life between my thumb and forefinger, and then we clicked off the light again and returned to our precious sleep.

On the night of the full August moon, we were awakened by the telephone. Ron, our neighbor who was captain of the volunteer rescue squad, had been coming home from a midnight ambulance run and found our horses—all nine of them—on the road in front of the farm. They’d been pastured nearby, on Ron’s side, next to his pond. Maybe someone had forgotten to turn on the electric fence, or else they had run out of good grass and smelled the clover on the other side of the road and decided to brave it. They had been in limbo when Ron came along, lingering around the yellow lines, not sure which direction to go. Luckily, it was a bright night, and no drunken or speeding drivers had been past before

Ron. He had hazed them into the field on our side before calling us. I pulled on my maternity pants and ran downstairs as quickly as I could, eight months pregnant. Mark bicycled to the barn to get the halters and lead ropes.

I could see all nine of them clearly in the moonlight. They had found the best patch of clover between two sections of the vegetable fields, and they had their heads down, grazing quickly and with purpose, like children bolting some forbidden sugar cereal. They were free, and they knew it. The clover was thick and almost to my knees, and walking to them was like wading in a shallow green sea. Tim's dappled gray team, the youngest and most playful, were the first to come to us. They were like big puppies, friendly and affectionate. They weren't wise yet to the connection between us and the loss of their freedom and that clover. They put their noses in their halters, trusting. Jay, one of the old half-Suffolk geldings, and Jake, the young Belgian, came too, and Mark walked that brace of four to the barn while I worked on the mares, Chad's team, and old Jack, the other Suffolk cross.

There is a sense of fun in horses when they know they have the upper hand. Abby let me get close to her, keeping her head down and still eating but with one eye and one ear on me. The last bite was taken with her mouth at an angle, because her legs were already moving away.

She stopped just out of my reach and put her head down again. I tromped after her with the big halter looped over my arm. We were still at it when Mark got back from the barn, the horse now clearly laughing at me. Mark and I worked as a team, and after an hour, all of them had consented to being caught except old Jack, who was just plain stubborn about it. In the end, he followed along free behind the others, walked into the barn, and went straight to his stall.

It was not the way I would have chosen to spend the night, but I wasn't sorry it had happened. How else would I ever have known how a herd of horses looks casting moonshadows into heavy clover that is silver with dew? And it was a treat to get to work alone with Mark. We had fallen in love over work, had poured all the energy of the early days of our marriage into it. Getting bigger meant we'd begun to lose the intimacy we had when we created the farm. It was a difficult adjustment. Our jobs had been cut up and redistributed among so many other hands. I'd begun to miss the shoulder-to-shoulder effort that had brought us together. We stopped at the coop on the way back to the house and filled our pockets with eggs. At the herb garden, I cut handfuls of parsley and dill.

It was too late to go back to sleep, so I put the water on for coffee and pulled the eggs from my pocket and started cracking. I love the dark morning time before the farm is awake. We moved quietly, talking low, so we wouldn't disturb Jane. Mark chopped dill, parsley, and thyme and sautéed diced onion with zucchini and a little bit of pork sausage. I cracked eight eggs into a bowl and sprinkled them with salt and pepper. They were pullet eggs, petite, and the yolks were bright yellow from the grass and clover the birds were eating.

We sat next to each other, facing the east window, and ate our breakfast. As the baby kicked inside me, we watched the sun come brilliantly up over the fields to the east, coloring the high clouds pink and orange and purple.

CHAPTER 5

My children's births foretold their natures. Jane's had been a long, thoughtful birth. A week before she was due, the rows of sunflowers we'd planted began to bloom. I took long walks in that field, between the nodding flowers that stood five feet over my head. Then Mark and I went to visit the hospital where I was scheduled to deliver. No, my body said simply as I walked in the door and smelled the hospital smells. No, it said in the elevator going up, sharing the ride with a groaning man on a gurney. No, on the maternity floor, where a red-nosed nurse was rude to us. I had the emotional hypersensitivity of pregnancy, and standing in the little room crowded with equipment, I said to Mark, "I'm not giving birth here." It wasn't a logical statement but a visceral one, and definitive. I was totally sure. "Okay," said Mark, taking my hand.

I knew other people who'd had their babies at home, with midwives. Mark's sister had done it twice. I had read credible studies showing home birth to be at least as safe as hospital birth for low-risk women. Earlier in my pregnancy, I had even looked for a midwife. The licensing laws in New York make it difficult to give birth at home, and there are very few midwives who do it. I couldn't find anyone close enough in our corner of the state. One of our members, a gentle, skilled woman who lived an hour north of us, was licensed for home birth in Vermont but not New York. Vermont was only a ferry ride away. For a while, we'd considered renting a room there, or borrowing an RV, but either of those options seemed at least as uncomfortable in practice as the hospital, and logistically tricky.

When we got home from the hospital, Mark made one last phone call, to a midwife we'd called before, and begged. She just couldn't do it. We lived too far away, and her other clients needed her. But just before she hung up, she remembered that a midwife she knew had just come back from living abroad and was practicing again. She put us in touch, and that is how it came to be that Jane was born in the same house where she was conceived.

I'd had an exquisitely healthy and active pregnancy. I had farmed gently but constantly, enjoying the alien feeling of my changing body, the sound of my own thickened pulse in my ears. I had to give up milking cows late in the third trimester, because I could no longer reach the teats over my belly. I stopped driving horses then too, because the bounce and rattle of the implements over rough ground was uncomfortable. But I could still move animals to fresh pasture, hand-weed, and walk the farm with Mark in the evenings. Between exercise, our own good food, and no alcohol, I was in the best shape of my life. I felt like a sleek animal, an athlete trained for birth. As the baby grew bigger, the hard weight of my belly began pressing down instead of out. The bones of my pelvis creaked, the sinews loosened.

Then it was late August, a week after my due date. The farm felt hot and dry. From the sultry kitchen, where I was canning tomatoes, I watched Mark drive the team past the house toward the barn, dust rising behind them. I was craving raspberry-leaf tea, and I had learned to pay attention to my cravings during pregnancy. In the second trimester, I had been overcome one day by an urgent need for chicken soup. I didn't have any chickens in the freezer, but there were dozens of laying hens in the coop. I marched outside, hardly thinking, and into the hen's pasture. Inside the coop, I found a hen pecking at another bird's egg. I picked her up and turned her over. She had a dull-looking comb and a dry vent, signs that she was no longer laying. Moreover, egg eating is a capital offense. Once a hen has begun breaking eggs and discovering their delicious contents, she is not likely to stop, and she can teach the habit to her sisters. It doesn't work, on a farm, to keep an animal who consumes more than she gives. I took her outside and stroked her feathers, thanked her for her work, then put her head under my foot and pulled up on her body firmly to break her neck. I skinned and gutted her over the compost pile, saving her heart and liver. Fifteen minutes after my craving had struck, she was in the pot with carrots, onions, celery, and herbs, and the kitchen was filling with the smell of chicken soup. It is true that the hen, in pecking the egg, was responding to her own natural craving. Sometimes it comes down to who is bigger, who is stronger. If we were equal-size, I have no doubt she could take me, because pound for pound, hens are certainly fiercer than I am. They are omnivorous, and good hunters. I have watched chickens peck to death animals as large as bullfrogs and as fast as mice. It is the way I imagine a stoning to be, death by accumulation of small blows that are collectively brutal.

The craving for raspberry-leaf tea required less violence to fulfill. We had wild raspberries on the farm, but they were scattered and hard to reach through heavy stands of goldenrod and poison parsnip. Our neighbor had recently moved away, leaving a large patch of cultivated raspberries, and had given us permission to pick them. I helped Mark unharness and brush the sweaty horses and put them out to pasture. Mark took Abby, and I led Jake, who kept a respectful distance from my awkward body.

The neighbor's empty house stood on the north side of our farm. A path connected us, blocked by a locked page-wire gate. Mark climbed over first, and I followed him, hoisting my belly over the top of the gate, landing heavily on the other side. We gathered a grocery bag full of raspberry leaves, and on the way back over the gate, I felt a tiny pop inside of me, and a trickle of warm water ran down my leg.

But labor was slow-coming. I waited, took long walks, drank raspberry-leaf tea until my eyeballs floated. Finally, Jenna, our midwife, advised me to take a dose of castor oil. This would evacuate my bowels, she said, and sometimes it got labor moving. It worked, and late that night, as Mark read to me in bed, I felt the contractions become steadier and stronger. We called Jenna, who planned to arrive the next day unless we needed her sooner. The contractions continued, but gently enough that I could sleep.

Jenna pulled into the driveway the next morning with her equipment and a cake. "A birthday cake," she said, handing it over. She was a sturdy, confident woman, gray-haired, opinionated, sure in all her movements. She was married to an obstetrician. In the course of her long career, she had caught more than five thousand babies, in all sorts of circumstances, all over the world. She examined me and found that I was several centimeters dilated. She listened to the baby's heart. "Strong and beautiful," she said, tapping my belly reassuringly. She gave me a bitter-tasting tincture and told me to take a walk while she settled in. Walking would bring the weight and pressure of the baby's head against the opening bones of my pelvis and trigger stronger contractions. I held Mark's hand for balance and walked with my legs swinging wide around the large hard rock of the baby's head.

We covered several miles that morning. I paused once in a while to let a contraction break over me. I was not afraid of the contractions. They weren't terribly painful. They were interesting, and elusive. The more I wished for them, the longer the pause grew between them.

We hiked all over the farm, talking, stopping

occasionally to rest. We sat under a tree in Fireman's Field, collected daisies, ate wild blackberries that were growing in the hedgerow. In the sugarbush, I hung by my arms from a tree branch, the absurd belly in front of me, willing gravity to work. When we came back inside, I was eight centimeters dilated, and Jenna told me not to stray too far from the house. I wasn't feeling much like talking, but I did feel like cooking. Jenna and Mark left me alone, and I brought in ten pounds of cucumbers to make a big batch of bread-and-butter pickles. I sliced the cucumbers and swayed and hummed in the steam of the kettle, the stinging scent of vinegar, the warm smells of peppercorns, bay leaf, mustard seed.

Just as the pickles came out of the canner, things got serious. A contraction came that was intense enough to stop time; I held on to the corner of the table as it washed over me. It was four in the afternoon. I went up to our bedroom and asked Mark to fill the birthing pool we'd set up. Jenna checked me again and said I was fully dilated. I got in the pool as soon as it was full. When I did, the discomfort disappeared entirely, but so did the contractions, which were still not as strong nor as regular as they needed to be. "Push," said Jenna. "You are going to have to work for it."

This part was not nearly as fun as the first

stage. It was shouldering large rocks up a steep hill. The pushing felt wrong, as though it were injuring me. It was like sticking your finger into a whirring fan or pressing your hand into broken glass. The kind of thing that a body rebels against. "Push," she said. "Push." I slipped into myself, my consciousness shrinking to a single porthole that looked out on a foreign world in which I had little interest. In the deepest part of the work, as the room was growing dark, I spotted a sunflower that Mark had picked for me and brought to our room. Its kind, calm face became a companion and a guide.

Jane was born after dark that night. She was bundled against me, skin to skin. She nursed, and then she slept nestled between Mark and me, and we were three instead of two.

If Jane was a sunflower, sweet and calm, then Miranda was a dragonfly, darting in on the wing. She came just before she was due, three years after her sister. Jane was asleep in her crib down the hall. Mark was asleep next to me. What I felt was nothing at all like what I had felt before. My body had learned the steps the first time, and now it was ready to dance. When I called the midwife, she said I didn't sound much like a woman in labor. Maybe I wasn't. Maybe this was the prelude. I took a shower, lay down again in the guest bed, enjoyed the sensations that were running through me, listened as they became more insistent. Mark woke up and timed them, then called the midwives again, this time with some urgency. It was a long drive for them through the dark mountains. I heard a whimper from Jane's room and went in to find that she had wet her bed. I stripped her sheets between contractions, put down towels, made up the bed again, changed her into dry pajamas. Jane had not woken all the way and sank back to sleep, her hand curled at her neck. Then I told Mark to fill the birthing tub, that I needed it now. He did, and as I stepped into it, I realized he had walked out to the garden at some point that night and gathered an enormous armload of sweet Annie. The room, candlelit, was full of it and its soft, candied, childlike scent. This time, pushing was no chore but a pleasure in its reflexive, peristaltic, rhythmic insistence. It had its own timing, and there was no controlling it. Body became separate from self, something to observe from a small distance with quiet wonder. I could feel the baby's head making its way through me, the same path her sister had forged with such difficulty, open now and clear, a relative speedway. I was aware, dimly, that the midwives had not arrived, that Mark was on the phone with one of them, checking location, getting instruction. I felt the crown of the baby's head underwater with my fingers, the sensation of fire. Then the midwife arrived at the top of the stairs holding a travel mug of coffee. She stripped off her jacket and knelt, and five minutes later, the baby was born, surfacing like a large-headed, small-bodied fish, to rest, bloodied, blinking, on my bare chest. We were still connected as I walked, supported, from the tub to the bed. There is no *we* stronger than those few moments when two bodies are connected. It was four in the morning. Mark woke Jane and brought her to the room in her monkey pajamas. Jane was unsurprised, serene. She had been practicing with her scissors for weeks, so she knew exactly what to do as she clipped her sister free.

It took us a week to name her. We scrolled through the list of names we'd made before she was born, and none of them fit. Finally, we landed on a new one: Miranda, with its echo of Shakespearean magic, and a triplet beat like her quick little heart.

Nobody had warned me that changing the shape of our family would be a different kind of labor. It wasn't just the work of having an infant, the feeding and soothing that happened around the clock, tedious but sweet. It was adjusting to a new family structure. After Jane was born, in my memory, at least, we'd slipped easily into a new rhythm. One little baby between two grown adults on the farm was doable. Aside from the breastfeeding, we'd split the duties of caring for her more or less evenly. Jane had spent her first winter zipped into our thick jackets, riding in a sling, or napping in the barn with us at milking time, tucked warmly into a bushel box. When we added a fourth person to our family, the dynamics got more complicated. A toddler plus an infant equaled one full-time job, and instead of splitting it between us or even discussing it much, we seemed to assume that job was mine. I soon found it wasn't an easy one. An infant who sleeps in two-to-three-hour snatches is manageable if you can take turns and get an occasional nap. But an infant who sleeps in two-to-three-hour snatches, along with a toddler who has her own set of needs, in the care of one adult, is a different thing altogether. Exhaustion built. The farm had grown too, and demanded more of everyone.

After Miranda was born, Mark seemed to be touched by some kind of postpartum unease. He took a two-week paternity leave from the farm but spent most of it in bed, reading magazines and sleeping while I nursed the new baby and cooked meals and played with Jane. Maybe, I thought, he needed a rest. There was something else, hard to say, but so it is: a gap had opened between us, a space that hadn't been there before. It wasn't us plus a child anymore. Now I was on one side, with the children and their needs, and he was on the other, with the farm and all its work. The sense of common purpose that had been so strong between us was smaller. "What's wrong?" I asked him. "Nothing," he said. A long pause. Then, "It's just clear to me where I fall on your priority list now. And it isn't first or second." I couldn't say it wasn't true. How could it be? That seemed to rattle him. When the two weeks were over, he got up and went back to the fields, in command of what suddenly felt like *his* farm, his battalion of farmers, his cavalry of horses.

Fall shifted to winter, the fields drifted deep with snow, the deer yarded up in the woods. The temperatures dropped so low, zero felt balmy. From the kitchen window, I could see the horses in the pasture behind the house. The pony followed the draft horses to water, her belly leaving brush marks in the snow. They chipped at the edge of the icy pond with their sharp hooves, dunked their muzzles into the water, and lifted them, the whiskers on their chins white with frost.

Racey had gone back to Africa. She had a four-month job in Central African Republic, working as a consultant for the World Bank. A few weeks before she left, Nathan had pulled into the driveway in a Subaru stuffed full of apprentice farmers. They had all finished their summer jobs and were on a road trip, exploring farms that sounded interesting or that they thought had something to teach them. There had been a stream of cars and trucks just like it that year. Word had gotten around that we were producing a full diet, year-round, for our local community, and that we were powered by draft horses. Young farmers wanted to see what that looked like and how it was humanly possible. Usually, we would feed them and find a place for them to sleep in exchange for a few days or weeks of work. For our full-time farmers, these visits could be enlivening, or they could feel exhausting. That day, for Racey, it was the latter. She was wrapping up a long season and mentally preparing to depart for the other side of the world. The last thing she wanted was to get to know a new bunch of people. They'd need to be shown how to do *everything,* and they'd probably end up sleeping on the floor at the Yellow House. But some corner of her brain registered Nathan's smile, his kind and friendly demeanor.

Nathan and his friends stayed for the weekend, and Mark gave them the Herculean task of mucking out a calf pen in the West Barn that was four feet deep in bedding. Racey was milking that weekend, and when she walked past them—to bring in the cows, feed the calves, let the cows out, check the heifers—she noted how efficiently they were working, with Nathan in the lead. They were having a blast, laughing and talking as the pitchforks bit deeper and deeper into the pile. Nathan stripped as he went, hanging pieces of clothing on the gate. A pair of long johns, an undershirt, and then a shirt. Each time Racey walked by, the muck pile was smaller, and he was happier, dirtier, and more naked. She allowed herself to be a little bit impressed.

We took a farewell walk around the farm the day before she left. She held Jane's hand. I had Miranda, tiny, on my back. Racey and I had become close friends, and she had grown attached to the girls. Most afternoons, she'd pop her head into the house, ask Jane if she wanted to collect eggs. Jane would rush to shove her feet in her rubber boots and run out the door. I knew the energy and time it cost to take a little kid along on chores, and I appreciated it. When Racey came in to visit and found Miranda fussing, she'd scoop her up and walk her around until she cooed or slept. Most people without kids of their own wouldn't think to do that, but it was natural for Racey. She wanted kids, I knew, but she hadn't met the right man. "What do you think," she asked as we passed the solar panels, "of having a baby on my own?" She was in her thirties, facing the fact that she was not going to get any more fertile. I thought about what it might be like alone. The sleeplessness, the stress when they were sick, the long marathon of patience it took to raise them, and also the life-changing surge of love. I imagined what kind of mother Racey would be, and it was a very good one. "Better to do it on your own than not to do it at all," I said. We talked dreamily about what it might look like if she built a little house for herself here, became a partner in the farm, and had a child who could grow up alongside ours.

Nathan decided he wanted to spend a year working with us. He took over Racey's jobs with the animals and in the dairy, and moved into her old room in the Yellow House. When Racey and I Skyped, I could see her equatorial world, glaring sun and dust and a fine expat's apartment, and she could see mine: the chilly, grimy kitchen, the children bundled in their heavy winter clothes, the dark and shut-up house. We talked about her work in a country that was just beginning to come back from a devastating war, and about a French paratrooper she was vaguely dating. One night she asked, "How's Nathan working out?" I said, "He's good. A little nerdy, maybe, but really kind, smart, and very hardworking. He asked about you." She blinked, shrugged. "Nothing wrong with nerdy," she said.

Another new farmer arrived, Tobias. He sat across from Blaine at lunch and looked at her openly from behind his big scratched-up glasses. She scowled at him reflexively, then smiled the barest hint of a smile. Blaine was dating someone else, but the energy between her and Tobias built day by day. You could see the heat rising between them like waves in the air over a hot patch of road.

The relationships that developed on our farm that year were like the ones that happen in a war zone. There was no hiding your true self when you were working side by side for grueling hours in all weather, dripping with sweat and muck. The farm provided shared purpose and a mix of frustration and satisfaction, which encouraged camaraderie. The beauty of the landscape and the physical nature of the work seemed to open people's hearts. Sometimes it was a friendship that developed, sometimes a romance. When that happened, love sparked fast and burned out or else caught and transformed into commitment. This pair, I thought, could go either way.

It was too cold to be outside with a newborn. The wind would make her gasp and choke. I stayed inside the house, cooked big lunches, and got news of the farm around the table. There was no fieldwork for the horses in winter, but there were a lot of odd jobs. Chad used a team to cut and haul firewood to be bucked and split for sugaring. When Blaine shot a steer in the field, Tim hitched a team to haul it home. I worried that the horses would mind the shot or the smell of blood, but neither seemed to bother them much. Tobias rigged up a ramp and a chain so the horses could pull the steer up onto the wagon. A smooth process, no tractor required. When it snowed, Tim and Nathan hitched a team to a blade we bought at an auction, to plow the farm roads. It wasn't a perfect rig, but good enough to clear a path to haul a load of hay to the beef cattle.

All winter, Mark watched the weather forecast, waiting for warm windy days. When they came, he rushed through his work as fast as he could, then headed for the lake. He was too busy to get away from the farm during the growing and harvest seasons, so he made the most of the rest of the year. When a January thaw came, he got his chance. The south wind built all morning, bringing rain that melted the snow. By noon, it had become a steady howling beast. I stoked the woodstove and put a pot of black beans on top of it, with carrots, onions, celery root, and hunks of salt pork. When the wind is like that, I want to feel four walls around me and enjoy the petulant sound of it trying to get through the cracks. Mark, on the other hand, wants to be in it, on the lake. His medium for this communion is a Windsurfer, and most of his gear is thirty years old. He had told me early in our relationship that if he dies on the water, I should know that he died happy, doing exactly what he wanted to do. It was midafternoon when he put on his dry suit, with layers of winter clothes underneath. The temperature outside was 35 degrees, warm enough, barely, to keep the water from freezing on his sail. “What time will you be back?” I asked, as I always did.

“After dark,” he answered. Dark would be over us in an hour or so. “You shouldn’t worry until ten.”

“What then?” I asked.

“Then you could call 911.”

I filed that in my consciousness without dwelling too much on it.

Once, just after Jane was born, I did worry. It was a stormy late-summer evening, and he was gone for hours. A tree came down on the power line, and the electricity went out. I drove down to the lake at dusk, with the newborn in her car seat, and scanned the whitecaps with binoculars. When it was truly dark, I drove back to the farm and sat on the staircase with the little babe asleep in my arms, imagining what it would be like to raise her without him. When he came home finally, he was limping, his feet cut from hauling his gear barefoot out of shallow water choked with sharp-shelled zebra mussels. He’d been blown so far downwind, he’d had to hitchhike home. I upbraided him for making me worry. “Worry is your choice,” he said. “I’m always going to be like this.”

After that, I didn’t worry. It was like turning off a little signal in my heart. And it was necessary if I was going to love the person he authentically was. But you can’t extinguish worry without also extinguishing some of the tender aspect of love. The physics of the heart won't allow it. You have to loosen your attachment or else you'll suffer.

So that warm, windy day in January, I did not worry. I hardly thought of him, even though I could hear the loose tin on the pole barn's roof tearing off in the dusk. I ate dinner with the girls and did not worry. I put them in the bath and did not worry. The trick of it, I'd learned, was to remove the center of yourself from the circumstance over which you have no control. The worry becomes an external thing, an object to observe and not experience. I'd gotten good at hushing the litany of what-ifs, stilling the slideshow of potential disastrous outcomes.

He walked into the bathroom after their hair was washed, trailing sand, half numb. He had come the last half-mile through the crashing waves in the dark, holding to the mast, his sail tattered and useless. His thigh was locked into a terrific charley horse that would not relent for days. I could see only the middle of his face, because the edges were covered by his thick rubber hood, but the visible skin was red from cold and the sand that the wind had whipped into it as he struggled onto the shore. The face was smiling a squished but unstoppable smile. He was enormously happy. That was Mark. And that was how he would always be. Happiest clinging to the mast, just this side of disaster.

CHAPTER 6

By February, the farmhouse was making me feel claustrophobic. It was built in 1920, redecorated most recently in the 1970s, and caught by us in a state of arrested decline, pinned there on a downhill slope. In the years when the farm was out of production, the house had been rented to a series of young people. It was a party house then, the closets wired and plumbed for growing pot, a pole for dancing. There had been some unfortunate renovations before that—the front porch enclosed in an awkward-looking frame, the ceiling of the dining room covered with dull acoustic tile, some of the large original windows replaced with others that were small and cheap-looking and positioned too high for me to see out of. One of them was cracked when we arrived and had never been fixed. I'd hung a piece of stained glass over the crack to keep from feeling a sting of guilt when I glanced that way. But the house's bones were solid. There were even some traces of the handwrought beauty of its youth: finely milled, shellacked fir moldings around the windows and the doorframes, which had aged to a glowing amber; still-sound horsehair plaster walls; and a hard maple floor that revealed itself, covered in stains, when we pulled up the layers of carpet and linoleum that we'd wrecked completely during our first year of farming, when we used to butcher sides of beef in the kitchen and separate our cream in a hand-cranked machine bolted directly to that hard-suffering maple floor. All that action had moved to the butcher shop and milk house, but traces remained in the kitchen. There was a hole in the ceiling where the meat hook had hung, and a patch of dark wood around the ghost of the old separator.

To get to the farm office, you had to walk a long U from the mudroom, where the fiberglass insulation was coming down in neon yellow puffs, through a dark, tight hallway, then through the kitchen and the living room, which also served as our dining room. There was a long table against the office windows. Its surface resembled the contours of Mark's brain: a crowded, complex kind of order that only he could understand. It was piled high with folders, papers, odd bits of hardware, the hay-dusted contents of his end-of-day pockets. There were jars marked with obsolete labels that no longer corresponded to their contents.

If Mark was not in the field or the machine shop, he was in the office, which meant there was a constant stream of people walking through the house to ask him questions, or look for something, or use the phone. There were two office chairs, rescued from the dump, with slightly uneven legs, and two beat-up old filing cabinets, the drawers slightly off their tracks. People needed to get to that office, or the kitchen, or the bathroom. With them came mud and manure that coated the floors in the spring and dried, in winter, to dust. Everywhere, there was visual evidence of Mark's hardcore functional aesthetic. He'd written important phone numbers on the wall in black Sharpie and punched nails into the molding of the windows next to the woodstove for hanging wet clothes, camping equipment, or a pair of socks he wanted to wear later on.

The living room formed the house's beating heart. Walking through it was hard because an enormous table and all the mismatched chairs took up most of the floor space. The bedrooms upstairs were always cold in winter, and so they were usually empty except during sleep, which happened under goose-down comforters thick enough to flatten us to our beds through even the most exciting dreams. The second winter of Jane's life, she spent every night sleeping in a pink snowsuit, cozy against the cold. It was a solution to the problem of keeping blankets on a toddler that seemed perfectly reasonable to me, then and now.

To improve the circulation of heat, Mark had hacked a hole in the living room ceiling, above the woodstove. There was a piece of wood lattice roughly nailed over the hole, so nobody would fall through. When visitors came, their eyes were drawn to it, curious and a little horrified. Who chops a hole in his own house? Jane would use it as a porthole, lowering her stuffed animals through it on a noose.

The porthole was useful because there were no interior stairs to join the first and second floors. That most logical connection had been severed before we moved in, when the upstairs and downstairs were rented out as separate apartments. In order to get from the kitchen to the bedrooms, we had to go outside to the unheated mudroom, come about, and go in an adjacent door. In the dead of winter, this was a bracing moment. If my hand was wet from washing dishes, it would freeze fast to the doorknob on the way to bed.

There was no stairway up from the living room, but there was a stairway that led down to the cellar, which was full, from harvest until spring, of sacks of potatoes and dry beans; eight-hundred-pound crates of winter squash; mesh bags of onions and shallots; and pallets of jarred tomatoes, pickled beans, and quarts of lard. There was a stainless-steel rack of cheeses aging in the corner, their yellow-orange surfaces covered in various fragrant types of mold. We lived, in other words, over a great, wide mound of food. If all of humanity were trapped in their homes by some apocalyptic event, we would probably be the only survivors, emerging, fat, from the ashes. It was a secure kind of feeling, but it had a downside. When the house was shut tight against the winter cold, and the woodstove glowed hot to keep the pipes from freezing, the smell of our cellar wafted up through the floorboards, permeated our clothes and furniture and even our skin. It was an organic smell, half good and half awful, made up of damp earth, onions, cheese, and, as we moved toward spring, the underlying stink of decay. We lived so thoroughly in that smell that I would notice it only when I had been away for a few days. When I returned, it would hit me as soon as I opened the door, and I'd think, *We really ought to do something about that.*

After a few years of harboring us, the house had seemed to molt. The wallpaper began peeling from the walls, steamed off by large-scale cooking projects: rendering lard, simmering down whey for cheese, and finishing maple syrup on the woodstove. There were six layers of wallpaper, including a 1920s floral print and something synthetic from the 1970s, the color of poi, an awful washed-out purplish gray. Once, on impulse, I'd begun yanking strips off the walls, thinking that maybe a bit more destruction would make it look chic. I was wrong. It looked leprous. Eventually, the walls and floors came to an agreement with the furniture and curtains, all of it moving toward the same shade of dun, barely illuminated by the light coming through the undersize windows, which were always in need of washing. At night, harsh shadows were cast by a bare and flyspecked bulb. Because of that general dimness, any bright spots in our house seemed particularly vibrant. A bunch of flowers on the table, or one of Jane's crayon drawings, were beacons of color and light.

For the first few years of our marriage, the state of the house was acceptable to me. Mark and I were focused entirely on what happened outside of it, in the fields and pastures and barns. The house was just the place that received our bodies when they were too tired to do another hour of work. It was where we paused to warm up or dry off, and where we cooked and delighted in the food that we were growing out there in the field, which was the part of our lives that really mattered. Sometimes friends from my old life would come to visit, and I would see their shock as they took in our house. I'd follow their gaze to the hole in the ceiling or a flock of cluster flies beating themselves against the cracked windowpane. It didn't bother me much. I even took a small, perverse pride in it. Other people could be house-proud. We were food-proud, land-proud, work-proud, and that was more than enough.

But it began to look different to me after Jane was born. My mother came to visit once, when Jane was an infant. My lovely, kind mother, whose house was always immaculate. She loved me with her whole heart, but if possible, she might have loved this new little granddaughter just a tiny bit more. She stood in the stairway that didn't connect the two floors of our house, my baby in her arms, and said in a masterful compression of maternal guilt, "You owe your child a better life than this." It stung. And it raised that pesky question I still hadn't answered: What is a good life? Is a good life for me the same as a good life for my children? And what if we can't have both?

After Miranda was born, the way I thought of the house began to shift. I needed to soften it, to civilize it, and to separate it somehow from the farm, with its rowdy band of farmers, its chaos, and its dirt. I needed an actual home for us, and that would prove very hard to come by.

The short days rolled on. I was learning that a farm was much less fun when you are an observer instead of a participant. Without the work, it was just a small business, ragged around the edges and only tenuously viable. Anchored inside with the children, I felt like it was slipping away from me, and Mark with it. When the cold spell lifted, I returned to milking in the mornings, thinking maybe that work would put me right.

Miranda, asleep on her back, had to be roused to nurse before I left for the barn in the morning. The feeling of holding her so fresh from sleep was goodness. When she latched on to suck, my milk let down, a tingly, prickly sensation fortified with a slug of pure love. Nursing her was taking a lot out of me. All the softness of pregnancy around my face and hips was gone, melted away by the caloric pull of milk, accentuating the circles under my eyes. Still, I'd miss this when she was weaned, just as I already missed the way she had slept in her first weeks of life, her arms thrown straight up over her head in a gesture of trusting abandon.

I could feel a bruise on my back against the hard wood of the rocking chair. The day before, one of the cows had kicked me there as I'd bent down to clean her udder. I'd been kicked many times before, in the knee, the middle of the thigh, the hand, the ankle. That was the first time I'd been kicked in the back. I hadn't known it was possible. But I couldn't begrudge her the gesture. We rub the udder to encourage the flow of oxytocin that enables the cow's milk to let down. The mechanics of lactation are no different for cows than they are for women, so I imagined the feeling of letting down was the same for her as it was for me. Maybe she kicked because what came at her then were my mute human hands and not the lowing mouth of a calf.

Miranda's sucking slowed to an occasional twitch. She was asleep again. I put her on her back and nudged Mark, left her with him, a quiet little bundle in a basket next to our bed. I slipped into the bathroom, where I'd left my clothes the night before. Layers upon layers, since the weather report had been calling for a deep freeze. Indeed, when I looked at the thermometer, it was twelve below.

Cold makes air seem cleaner, clearer. Shards of moonlight glinted off the snow, and my boots made a high squeak with every step. I was armed with a butane torch, a can of kerosene, and my hair dryer, in case the pipes in the milk house were frozen. In the barn, the fluorescent light struggled weakly against the cold. Nathan was at the door with his stick, calling cows: “Come on, come on.” I joined him, shining my headlamp at horns and undercarriages, looking for Steve, the new bull. He came along behind Cammie, who was in heat. “Hey hey hey,” Nathan said, and shook his stick at him. The bull turned in to the bullpen, where a pile of hay with a sprinkle of corn was waiting for him, and I locked the heavy gate behind him. The last cows filed in. We followed them back to their stanchions, where they’d found their places and were unfurling their tongues for the first tastes of corn.

We set the kerosene heater in the aisle of the barn and ran it full blast. It threw off a paltry heat that dissipated within a few inches. I went to the milk house to fill buckets with hot soapy water for teat cleaning. Two minutes after leaving the heated space, the suds on top of the water were frozen into a brittle white foam.

I’d gotten to know Nathan better by then. He was from Massachusetts, like Racey was. He had a degree in geology, spoke fluent Spanish, was a masterful tennis player, a good musician, and already knew that he really, really loved farming. His last job had been at a small dairy in Canada, where he’d learned to drive draft horses. When he came to our farm, he improved so many systems so quickly that we barely knew where to put him next.

When I reported all that to Racey, in Africa,

over Skype, she said, "Man, is there anything that guy can't do?" I knew he took an interest in her, developed during the two weeks their jobs had overlapped. I'd overheard him talking with Tim about her in the milk house. Nathan sounded thoughtful, sincere, and determined. That might be a dangerous tone to take with Racey, I thought. She tended toward roguish men who lived on other continents and offered excitement in place of intimacy. But maybe.

I hadn't milked regularly since before I got too pregnant with Miranda. The milking machines had changed the nature of the work, and it was new to me. Instead of the contemplative crouch and slow squirt-by-squirt action, milking had become a solitary aerobic activity, one person, two machines, fifteen cows. We had not come entirely into the modern world. The milking machines we were using were from the 1960s, five-gallon sealed stainless-steel buckets that we could pick up and move from cow to cow. From the top of each bucket sprouted three lines: one for the milk, one to provide suction, and the third to feed a genius little plastic-and-metal regulator that allowed the suction on the teats to come and relax in one-second pulses, much in the same way a calf would do it. The bucket milker's line attached to a suction pipe that ran in front of the cows' stanchions. The business end of the bucket milker was the claw, which had four rubber tubes inside stainless-steel covers. Milking meant first cleaning the teats and udder thoroughly, which would also stimulate letdown, squirting the first milk from each teat by hand, then attaching the claw. When it was finished, we took off the claw, sanitized the teats and the machine parts, then worked through the same process with the next cow in the lineup.

Usually, two cows' worth of milk could fit in each bucket. When it was full, I would take off the bucket's top and empty the milk through a filter into a ten-gallon stainless-steel can. Once I learned the timing, it was an enjoyable process, the pace brisk but manageable. Even on the coldest mornings, it warmed me enough to take off at least one layer. By hand, milking took us about ten minutes per cow. With the machine, we could milk two cows at once in as little as two or three minutes. We lost the animal-to-animal intimacy of hand-milking, and the congenial morning and evening talk between milking stools, but I soon came to think the cows preferred the machines for their speedy efficiency and the reduced wear and tear on the teats. Anyway, we needed the increased efficiency so we could stay on top of the farm's fast growth.

The first time I'd milked after having Miranda, I'd been all thumbs, awkward. The routine jobs—

chores, harnessing, milking—are like dances. The first time, you trip over yourself. It takes forever, maddening for the learner and a drag for the teacher. A few days later, though, you know the steps. Then you can do it without thinking at all, and without wasting any movement, and lose yourself in the flow, your mind quiet or noodling on some idea or noticing important things at the edge of your vision—is that cow looking a little off? Is she coming into heat? Meanwhile, your body is in motion, at medium speed, over long periods of time, releasing puffs of endorphins that leave you happy.

Cold days are always more complicated, especially with milking. Liquids become problematic solids. Buckets freeze to the floor, hoses clog with ice, doors stick, and stanchions jam and are difficult to fix with numb fingers. If you leave a drop of milk or the liquid iodine dip on the tips of a cow's teats, she'll get frostbite there, in the worst possible place. At fifteen below, the pulsator slowed to an ineffectual *swish-swish,* the valve to the suction pump jammed, and the water line in the milk house froze. By the time all the problems were solved, the cows milked and turned back out for the day with Brian the bull, it was nearly midmorning, and my own breasts were full to bursting again. I hustled back to the house, where I found Mark in the kitchen, talking through the day's work plan with Chad, while holding a fussy, hungry baby and corralling an equally hungry toddler. I gathered the children, took a plate of leftover breakfast, and trooped upstairs with them, shedding layers as we went.

That night, a snowstorm hit. Not a normal snow-globe snowstorm but the quiet, vicious kind that comes every few years and drops snow so heavy and wet that it tears down the power lines, the fences. Two barns in the neighborhood collapsed under the weight. Every hour or two through the night, we went outside to gently coax the snow from on top of the greenhouses. We were lucky. The barns stood, the house stood, the greenhouses stood. In the pasture north of the West Barn, the dairy cows stood in a circle around their hay feeder, then came in for milking one foot at a time, their full udders leaving drag marks in the snow. The chickens stood at the barn door and pecked at the snow but would not step out, preferring to mill together in the warm, dry barn. The pigs had feed and shelter in the field. But the beef cattle were pastured a mile from the barnyard, in Essex Field. They had enough hay for twelve hours, maybe eighteen, but then they would be hungry, and the electric fence that held them in had no power. The last time the cattle had been hungry inside a compromised fence, they had walked the dirt road toward town, where we'd found them nibbling the shrubs around the statue of the Holy Family at St. Joseph's Church. So we needed to feed them, but their hay was here, in the barnyard.

The roads were impassable. This snow was too heavy to be plowed with a blade and had to be scooped out of the way slowly, with the bucket of a tractor. We were on our own, and there was so much to do. We split up. After milking, Mark took the tractor and began making his way, bucket by bucket, north of the barns. I took the children and the barnyard chores. The snow was too deep for Jane to navigate, so I put her and the baby on a toboggan and pulled them toward the East Barn. Jane was in a heavy snowsuit, and Miranda was practically immobilized by extreme bundling, with just her pink face poking out. She fell asleep almost as soon as the sled started moving. By the time I had trudged, thigh-deep, to the East Barn, I was winded. I needed to go in and feed the chickens and collect the eggs. Jane was happily humming to herself, and Miranda was sound asleep, and I could do it faster without them. "Stay here," I told Jane, "and watch your sister. I'll be right back."

I collected the eggs as fast as I could, made sure the chickens had enough feed and water to get through until the next day. As soon as I opened the door of the barn, I heard Jane wailing. Not fussing or crying but the most brokenhearted wail, as if someone had died. "Mom!" she bawled, unable to get her breath. "Mom! There's a cat! On my sister!" Indeed, there was Bubby the barn cat, lying on Miranda's cushioned chest, energetically kneading her claws into the red snowsuit, staring at the baby's face. Miranda was lying like a starfish on a bright white beach, still peacefully asleep. To me, the scene was innocent, even adorable. To Jane, it was terrifying, awful. The three rules we had taught her when her sister was born were these: "Always be gentle. Don't put anything over her face. Don't put your weight on her." The cat was putting its weight on her, and to Jane, this was going to kill her, and my instruction had been "Watch your sister." Jane had tried to swat Bubby away, but the scarred old battle-ax of a barn cat knew a good thing when she found it, warm, soft, and dry.

There were times when we asked too much of our children, Jane especially. She had the innate sense of duty and responsibility that I think is common to oldest siblings, whether they are born on a farm or not. Once, when Miranda was mobile but not old enough to understand boundaries, we let Jane take her to the swing set at the side of the house. "Don't let her go out on the farm road," we said. But Miranda was willful, and I found Jane, hysterical, trying with all her four-year-old might to haul the toddler back from the road, where there was no chance of her getting run down except in Jane's imagination. But for her, it could not be more real.

And there were times when I asked too much of myself. One late-winter weekend, Miranda and Jane both had colds, and nobody had slept in two nights for all the hacking. But it was okay because it was Sunday, and I was going to sleep in because we had Dylan.

Dylan was a wiry, energetic young farmer from the next town who had started his own vegetable business the year before. He worked insane hours during the growing season, employing a shifting cast of Jamaicans and hippies, all fueled by quarts of black coffee. They often harvested by headlamp at midnight, blasting jam-band music out of big speakers hung on the rafters of the packing house. They had installed a tap through the wall of their walk-in cooler, stocked with kegs of beer. Dylan's farm was a good place to go at any hour if you wanted company, work, or a party. But sometimes the partying made Dylan less than perfectly reliable.

After frost, Dylan's workload lifted because he wasn't raising stock. He usually spent winters in Ecuador, where he bought and sold coffee and hung around in sunny hammocks, but this year, he and his girlfriend had had a baby, so he was staying home and was available for short-term hire. He had worked on a dairy before starting his own farm, so he knew how to milk and care for cows. I had been frayed at the edges for weeks, and I knew without fully acknowledging it that the first serious cracks were beginning to appear in my ability to cope. As a way to forestall actual breakage, I'd hired Dylan to milk for me on Sunday mornings. That first beautiful morning had finally arrived. To say I'd been looking forward to it is insufficient. I'd been clinging to it like a waterlogged sailor to a bit of flotsam. I'd gone to sleep knowing that Dylan would milk, Mark would do the rest of the chores, and I and the two kids would snuggle back in for a little more dear, dear sleep.

From under a thick pillow well before dawn, I heard Mark get up and dress, trot down the stairs, and walk out the door to the barn. Then, with growing dread, I heard the disconcerting noise of all of that happening again, but in reverse. And there he was at the doorway, rousing me as gently as possible to tell me that the cows were not in the barn, that Dylan had not shown up, and that I needed to go milk the cows.

What was I going to do? Mark had the morning chores to do, and they couldn't wait. The cows had to be milked. That couldn't wait either. There was no use complaining. I pulled the runny-nosed children from their beds, waking them in the process. They whimpered at the injustice and resisted my attempts to put four arms in snowsuits, four feet in socks. When everyone was dressed, Jane kicked off her boots and declared that she needed breakfast before going out. That, I realized, was a reasonable demand. I buttered a piece of bread for her, poured myself a cup of cold coffee from yesterday's pot, then looked at the clock and hustled. Hustling sick, tired children is the worst idea in the world. There were now tears from both small sets of eyes, and two pitches of squalling. But finally, they were both in the stroller, bread in one hand, mittens on all three others, and we were rolling.

We got to the barn and let the cows into their stanchions. Their dishy faces ranged in color from light fawn to nearly black. Zea was the new girl, a nervous, jumpy cow who had just had her first calf. Delia was gone. Her daughter Sis had taken her place in the stanchion by the door.

Jane climbed out of the stroller to watch me assemble the milking machine. I ran buckets of hot water to wash the cows' udders, and put the ten-gallon milk cans and the big stainless-steel filter in the alley of the barn. Then I recruited Jane to push the baby around in the stroller, a tactic meant to occupy them both.

I got the first cow's udder washed and prepped and had just put the milking machine on her teats when the stroller rolled into the gutter at the other end of the barn and dumped the baby. As I ran to rescue her, the cow kicked off the machine. I tried again, employing the cats this time to entertain Jane, parking the crying baby in front of the milker, where I hoped the *suck-suck* noise would soothe her. When one of the cats scratched Jane, I dumped five gallons of milk into an already full milk can, and half a cow's daily production ran onto the barn floor. Jane wailed over her scratched hand. Miranda wailed over the cold wet muck on her snowsuit. I wailed for Mark in a voice that implied it was all his fault. He left his chores, noted the milk in the gutter, scooped up both kids, took them to the milk house, and plopped Miranda on the floor, where I could hear her crying herself to sleep. That was when Zea took advantage of my inattention and kicked me, landing a good one on the kneecap. Silent rage bubbled out of me, not toward the cow but toward Dylan.

I spied the cup of cold black coffee on the step to the hayloft. Caffeine might substantially improve my attitude, which was spiraling fast. I milked Sis, the last cow in the lineup and my favorite, inheritor of Delia's sweet nature. Over the noise of the milker, I could hear Mark singing with Jane in the milk house as he washed the pails—"cockles and mussels alive, alive, oh!"—and Jane correcting her dad on the lyrics. Miranda was finally quiet, asleep. I took the claw off Sis and put my coffee cup under her. There was still some good hindmilk in the back teats, and I grasped one, warm and soft in my hand, and stripped it out. The rich milk hit the coffee and foamed, a farmer cappuccino. I milked until my coffee was both light and warm. I fed the cows and sipped my coffee. Then I stood still and listened to the sounds of the cows munching and my family singing.

This was not at all what I had pictured, way back at the beginning, when I imagined raising children on a farm. How would they think of their childhood when they were my age? Would the stories they told be about pretty cows or tears? Maybe it hinged on whether Mark and I could figure out how to navigate the dangerous territory we had entered. As the winter wore on, we were increasingly in opposition. It was more than sleep deprivation. It was not just the change in group dynamics. It was a long hard game of family versus farm. Which needs would get met when there wasn't sufficient time, money, or sleep? I was looking at the present and taking care of the children. Their needs were more important to me than the farm's. Mark was looking at the bigger picture and working on behalf of the farm first, thinking long-term, even if it meant sacrifice or unhappiness in the present. Two strong-willed people pulling in different directions created a lot of tension.

• • •

The winter weeks wore on. Milking was liberation from the house, but it added to my exhaustion. On nights when Miranda didn't sleep well, it felt nearly impossible to get out of bed at four-thirty and continue all day. The house, meanwhile, seemed to be closing in on us, filling with visiting young farmers interested in our horses and our full-diet model. Mark loved company, new opinions, and the constant underlying hum of youth and action, which fueled our winter work. To me, the house felt increasingly crowded and dirty, and there was way too much noise for a family with an infant. Someone would bang a pot too hard against the stove or scrape back a chair in front of the woodstove, and the sound would rattle the baby from her hard-won sleep and shave away my last nerve. Or a salty word would land in Jane's receptive ear and become part of her three-year-old vocabulary. Or the work of cleaning up after so many people would overwhelm me. Mark came up with what he thought was the perfect solution: he assigned everyone his or her own place setting, and everyone could choose to wash or not wash their dishes before the next meal. One afternoon in the pantry, looking for a cup for Jane, I picked up someone's proprietary mug and found it stuck fast to the plate beneath it.

I called my friend Emily, who lived a few miles

up the road and had two boys who were a little older than our girls. "Tell me it gets easier," I said. "Or if it doesn't, just lie." She came over, washed the mountain of dishes in the sink, and left dinner for us, and a pie. The next week, she drove down the hill again, after dark, in the middle of another snowstorm. She had her skates and a cooler with two bottles of beer. The cooler was not to keep the beer cold but to keep it from freezing.

I hauled my own skates out of the garage and pulled on my ski pants and followed her to the pond. The snow was falling fast. We each had a shovel, and we scraped away at the inch of fresh snow on the ice until we had made a loop. By then we were warm in our thick clothes. Flying around the edge of the pond, without the weight of a child or chores, felt impossibly fast and free. When we were tired, we hunkered down into the snowbank at the northern edge, under the shelter of a clump of sumacs, and opened our beers. The falling snow made a sort of private room out of the open air. We talked about children, and me turning forty, which I would do in a few short weeks, and I vented about the rapidly changing farm and my shifting marriage, both completing their itchy seventh years. I told her about the long nights, my exhaustion. She took it all in and nodded. She never did say it would get better. Neither did she say it wouldn't.

Instead, she quoted Wendell Berry, a line from a poem that I had known well once but forgotten. " 'Be joyful,' " she said, " 'though you have considered all the facts.' " It made me laugh, and gave me hope. Maybe this time of year was not for avoiding the dark and difficulty but for embracing it, finding the hush in its depths.

The next day, I walked to feed the beef herd through a four-inch layer of perfectly fluffy snow with Miranda asleep on my back. The storm had passed, and it was cold and crisp and the sky was blue, mottled with dense clouds that sent forth slow fat flakes. Down the hill next to Long Pasture, I walked past the cabin built by one of our first employees as his housing, before we figured out that it was better for everyone when employees had a place away from the farm to go home to at night. My eye caught something wrong. A hole. The cabin's window was shattered. In my exhaustion, I thought, *Vandals? Here?* I swung open the door. The shards of window glass covered the mattress, the floor. The hole was enormous. Then I saw the bird. It was curled belly-up on the floor, a drop of blood next to its beak. I picked it up, stroked its cold feathers. It was large and heavy for a bird. What was it? Quail? Pheasant? My mind flicked through the birds I knew. Then I got it. Grouse. Not just grouse but one particular Grouse, the one that always made a startled noise as I walked down this hill, then flew in front of me across the farm road, at eye height, into the woods around the cabin. I guess the last time she had an unlucky trajectory. Fear is a terrible accelerant. *A flight of panic ends in shards,* I thought. *Remember that.*

Mark met me in the field, driving Jay and Jack hitched to a wagonload of hay. Jane was perched next to him, holding on to the bale strings. I opened the gate for them and settled the baby next to her, safe between two bales. The herd of cattle came lowing through the snow. Mark drove the horses and tossed the bales off in a line, and I followed along, cutting them open, collecting the strings. With an ax, I chopped a fresh hole through the ice of the spring where the cattle drank. Then I jumped back on, and we drove up the hill toward home. Going past the cabin, I felt my heart tighten. I knew all of a sudden what we had to do. It wasn't a thought—my brain was too tired for that—but a certainty. "We need to get the farmers out of the house," I said flatly. This was an argument we'd had many times before. Usually, Mark argued for the efficiency of one person cooking for the whole group—he had calculated it into minutes per meal served—and in winter, the efficiency of a single heated space. But this time, he looked at my tired face, the downward shape of my mouth. "We can make it happen tomorrow," he said.

• • •

We called a farm meeting and broke it to them. They could eat breakfast at the Yellow House, before morning meeting, and go back home for lunch or bring it along with them. No more huddles around the woodstove. The office was moving out of the house too. Mark got on the phone and found a beat-up old trailer that had been the on-site office for a construction company. We had an electric space heater for it, and a stove we could install for cooking. Until then, the farmers could warm up during the day in the greenhouse. We'd continue to host dinner every Friday night at the house for anyone who wanted to come. But during the rest of the week, the house would be for family.

It didn't go over well. Chad didn't say anything. Nathan and Tim nodded, quiet. Tobias looked at Blaine. Blaine looked angry. I could understand why. On our farm, meals were an important part of the deal. The work was hard and not well paid, but the meals we ate together were amazing, which, for some types, helped the job make sense. The food, after all, was why we were here, all of us. "But the table is where the magic happens," Blaine said, shaking her head at us. And she was right, but so was I.

On Valentine's Day, we got a babysitter, and Mark and I spent the afternoon working together, a farmer date. We'd harvested the last of our corn by hand, and stored it on the cob, and now we needed to get it shelled. We'd borrowed a combine and parked it in the barnyard. I sat in the cab of the John Deere and worked the levers while Mark dumped loads of corn into the maw. It was something, to sit high up in that giant machine and watch the cobs go in the front, the clean kernels come out the augur in the back. We were three-quarters finished when pieces of cob jammed up somewhere deep in the bowels of the machine, and the combine choked and died. Date over. I went to collect the girls, and Mark, Nathan, and Tim spent the rest of the day underneath the combine, trying to free things up. Mark came back to the house at dusk, luminous with corn dust, like a coal miner in negative.

Blaine wasn't hiding her feelings about getting kicked out of the farmhouse and took a leave from the farm, traveled to Guatemala. We had less work in the winter, and she'd be back in spring. It felt easier with her gone. The tension between her and Tobias was palpable, but nobody knew where they stood, and I didn't dare ask. Two men had given her rifles for her birthday; Tobias had had the wooden stock of his inscribed with her name. Having her gone for a while would make things smoother. Everyone else accepted the change without comment. But it was a line between us and them. I didn't quite understand that lines between us and the people who worked here were necessary, not a bad thing. Back then, lines made me uncomfortable. And there were starting to be a lot of them. There was the line between us the farm owners, and they the farmworkers, which was thin when it came to the type and intensity of work we did but got thick when there were disagreements, or when someone started to wonder why they were working so hard, anyway—and thicker still as they gained skills, began dreaming of their own land and how they'd like to work it. Increasingly, there was age. When we first started hiring people, our farmworkers had been our contemporaries or close to it. As we got older, the rotating cast of people we hired did not. They were all somewhere between young-young and middle-young, and I'd just turned forty, which made me officially old-young. There would be lots of good, even lifelong relationships to come, but Racey would be the last employee who felt like my friend.

The biggest line of all was having children. I knew how hard it was for them to imagine what it was really like with kids, just as it had been hard for me to imagine before I had them. Once, after Miranda was born, I'd come downstairs to join a table full of farmers for lunch. The front of my sweatshirt was stained with breast milk, the baby was crying, and Jane was taking food from my plate and then wiping her hands carefully in the baby's hair as I tried to hurriedly eat. I heard a guy who was visiting the farm for a week—a young kid, still in college—lean over to his friend as he glanced at me and my armload of children and ask, "Why on *earth* would anyone sign up for that?" He didn't mean the farm or the crowded house. He meant having kids. So the farmers, all young, single, and childless, didn't really understand why we were kicking them out of the house, and they didn't like it, but for the most part, they accepted it. The rest of the winter, I'd walk past them at lunchtime, eating together miserably in the dripping greenhouse, and feel a strong mix of guilt and relief.

The house was so quiet without the farmers in it. I took all the extra leaves out of the dining table and shrank it from an enormous rectangle to a four-person square. I made small delicious lunches for the girls, unused to cooking at a family scale, and the three of us ate clustered together around one corner. We would set a place for Mark, but most days, he didn't have time to come in. As winter faded, the separation increased. Between me and Mark, firmly on different teams, with our opposing needs. Between me and the farm and its endless stream of demands.

PART 2

BEDROCK

CHAPTER 7

March arrived, sweet and wretched. The barnyard was four inches deep in mud, the beef herd had its annual case of mites, and the ground itself looked patchy and worn, bare land showing through snow in circles under each tree. Then the south wind brought the smell of spring. We'd tapped the maple trees during the last cold stretch, when it seemed like winter would never break. On the first Sunday, the thermometer hit fifty before noon. The forecast was perfect for sugaring. A freeze was predicted for that night, and another warm day the next, with the same pattern of warm days and cold nights likely to continue through the week.

Sugar season brought some relief from the claustrophobia of the house. Sugaring, the way we do it—tapping the maple trees by hand, hanging them with buckets, and collecting the sap with the horses—is joyful work, made to be done with little kids. Because the first run was on a Sunday, the farm was quiet after the morning milking and chores, just us family. Mark and I brushed Jake and Abby, then harnessed and hitched them to the sap wagon while the children took their naps. In the late afternoon, Mark and the horses picked us up at the house, the girls bundled into layers, and we set out for the sugarbush together. Miranda rode on my back and Jane ran through the woods from tap to tap, catching the drips of sweet sap on her tongue. When her legs got tired, she hopped on the back of the wagon and rode. I carried the full buckets of cold, clear sap to the wagon and dumped them through the filter, which caught the stray pieces of bark, the moths that had perished in the sap overnight. Jet trotted along behind us. This was how I had imagined it when I thought about raising kids on a farm. I'd almost forgotten how happy it made me to work together outside in the fresh air. It felt good to stretch my legs, tromp through the mud, the last piles of snow, in the cold air and sunshine, and feel the woods coming to life around us. Mark whistled to himself at the front of the wagon, lines in his hands. On the straight flat path at the top of the sugarbush hill, the horses picked up their feet and trotted, arching their necks, caught up in the good spirits of the day.

By the time we had collected all three hundred buckets, we and the horses were sweaty and muddy, and night had begun to fall, bringing a chill that would freeze the trees and send the sap back to the roots, to rise again the next day. We needed to begin boiling so we would have room in our tank for tomorrow's haul. We backed the sap wagon into the pavilion, near the evaporator. There was a stack of dry wood next to it and a holding tank above it, suspended by an enormous set of chains. We pumped the sap up into it, bucketing the last of it by hand. It was past the children's bedtime, and Miranda was sound asleep in a little nest Mark had made for her, under the eaves but away from the sparks, encased in her layers of fleece. Jane was thrilled to be working with grown-ups, around all that sugar, up late in the dark. She played outside the pavilion in the mud, brought twigs to burn, stomped at the ice forming around the edge of the stream. Then I took the horses to the barn to brush and feed them while Mark gathered kindling, stacked it neatly in the evaporator, and put a match to it, igniting the sugar season.

By the time I got back to the pavilion, the main

pan and the finishing pan were both at a rapid boil, and the air was fogged with steam and smoke. The first boil of the year is the one that sweetens the pan. It takes a long time because you are starting from scratch—all sap, not syrup—and must drive off almost all the water.

We ate a supper of eggs that we'd hard-boiled in the sap of the finishing pan, piling the sticky shells on the ground. Mark and I took turns stoking the fire, adjusting the valves and floats.

We drew the first syrup at midnight. By then Jane was asleep next to her sister, cuddled into a sleeping bag that I'd fetched from the house. She had been sleeping outdoors regularly since she was eighteen months old—summers in a tent because of mosquitoes, winters in heavy bags, under the stars—and she could fall asleep outside as easily as she could in her own bed.

The first draw is the clearest and lightest syrup, a pale delicate amber color with a pure, sugary taste. Later in the season, the finishing pan would hold more caramelized syrup, and the color would be darker and the taste more robust, but this was the fancy stuff, the precious first sweetness of the year. We got a good bottle of peaty Scotch from the house—a Christmas gift from my friends Nina and David—and toasted the coming season, the completion of our family, four people on this piece of good land, working together. It felt good, with a glass of Scotch in my hand, to sense the firm weight and certainty of that work and envision the years ahead. The cycles of the seasons, the tides of dearth and abundance, the turning of the calendar. It was exciting to feel that we had found our permanent form and could spend the next decades of our lives within its bounds, tending it.

I wasn't exactly right about that, of course. We were still actively wrestling that spring with the problem of scale. Scale was nothing and everything. We were growing the same diverse array of plants and animals we had our first year, to feed a group of people directly. But a change in scale meant a change in the type of equipment we used, the number of people we employed, the kind of infrastructure we needed. There's a limit to how big a farm can get without changing its nature. There's a limit to how small it can be and stay afloat, pay its taxes, its mortgage, provide a living of any sort for a family. We argued about this a lot. Mark saw the big picture. We couldn't stay small and diverse and survive economically. I saw the details. We depended so much on human labor that a bigger farm meant a more crowded farm, a more complicated farm. All those personalities and changing moods mixed together, sending out cacophonous emotional noise even when things were going well. And we were both insecure about investing in the expensive improvements we needed to make, because it meant taking on debt. Debt, on a business that still felt precarious, was scary.

So we inched our way toward larger scale that spring as cheaply as we could, using half-measures, trades, and bargains. Our basement wasn't big enough to store all the potatoes and squash we planned to harvest that year. The cold root cellars, in the foundations of the old round barn, were not reliable enough to keep our cabbage and carrots from freezing or rotting or both. Mark found another tractor-trailer box. The ground was frozen when he bought it, and it had to be ripped out of the hard mud it had sunk into, so the front end was torn and tattered. We parked it next to the butcher shop and added insulation outside, refrigeration inside. Mark saw how cost-effective this was and beamed at it every time he walked past. I saw its unremitting ugliness and averted my eyes.

Racey came home from Africa in the middle of sugar season. She'd had enough of communal living at the Yellow House, and rented an apartment a few doors down, in the servants' quarters of the mansion across from the ferry. She invited me over for a homecoming dinner. Over a bottle of red wine, I brought her up to date on the girls and the farm, and she told me about Africa. Her job had gone well, but the peace in Central

African Republic was looking fragile. She was glad to be back but worried about what might come next for her friends and colleagues there. We talked about the French paratrooper she'd been dating, Lolo, who was stationed in Bangui. She had gotten in a little too deep this time, for something that wasn't ever going to fit, but it felt safer now that they were on different continents. Lolo Skyped her in the early mornings, before she came to work at the farm. Also, Nathan had friended her on Facebook while she was away, and had sent her a string of letters. He was a good writer, she'd discovered, and insightful. I raised my eyebrow at her. "Nope," she said. "He's way too square for me."

Later that week, a late-season storm dropped two and a half feet of snow. It slid off the roof of the barn into heaps higher than my head. We had a boar we'd rented who needed to go back to his owner. Mark had assigned Racey and Nathan to dig a chute through the snow for him, from the barn door to the stock trailer, which was parked twenty feet away, as close as they could get. It was heavy snow, a big job, and it had taken the two of them all day. I could see them from the upstairs window of the house, where I was nursing Miranda, and could hear them talking and laughing while they worked, and then a hoot as they ran the boar successfully onto the trailer. They looked like they made a good team.

• • •

There followed a string of warm days and cold nights. Sugar season was long and plentiful. Smoke and sweet steam rose without stopping from the pavilion. Mark didn't trust anyone else to run the evaporator, so he spent long nights out there alone, then started his workday at dawn. His energy seemed bottomless and held the whole potential of the year. On the day of the last good run, the horses went to the woods with the sap tank three times, bringing it back down the hill each trip sloshingly full with two hundred gallons. Then the trees put an end to it. Their buds swelled and began to break. Sap from budded trees smells sour when it begins to boil, and the syrup tastes acrid, exactly as awful as the good stuff is good. We pulled the taps, thanked the trees, and put seventy-five gallons of syrup in jars for the year.

Our house was mostly free from farmers that spring, but not from farming. The first batch of chicks arrived, three hundred fluffy yellow balls of life that arrived in the mail, making a high-pitched racket in the post office, prompting Susie the postmistress to call us before she rolled up her window and say sweetly, "Please come pick up your box of noise."

Racey and I set up the chick brooder on the house's enclosed porch, the only space available that had electricity and was also dry and reasonably free of rats. The porch's drywall ceiling had melted from a leak in the roof before we arrived. The leak had been fixed, but the ceiling had not. As Racey filled the feeders, I mixed up a gallon of my special chick elixir—sap, salt, and a splash of cider vinegar in a gallon of water—meant to refresh them after their long trip and ease them into the delicate first days of life. The elixir was mostly superstition on my part, but we'd had good results with it. We dipped each little beak in it as they came out of the shipping box. The brooders were awkward but terribly efficient things that could accommodate 150 chicks at once. They were made of plywood, on short legs, with lightbulbs inside to make a warming box. The chicks could run underneath to get warm, then increase the length of their forays into the chilly world for food and water as they grew. When they could regulate their own body temperature and withstand some cold and rain, they could go to the pasture. Until then, they depended on us for survival. We bedded underneath and around them with fresh hay twice a day, refilled their waterers in the kitchen. No amount of sweeping and tidying could keep the smell of hot chick from rising through the hole in the porch ceiling to our bedrooms above.

We were also growing Belgian endive in the dining room. Endive gave us a taste of fresh leafy vegetables at the tail end of winter, a luxury, but growing it was a long-term, labor-intensive project. We'd planted it the previous spring, and it had spent the growing season developing a hard, fleshy corm—a rootlike structure, similar to a bulb, that stores the plant's energy underground—and its characteristic chicory-family leaves. In the fall, after the leaves died back, we'd dug up the long row of corms and buried them in plastic root bags filled with potting soil. The corms slept, dormant, in the cellar of the farmhouse through the winter. The last phase required the most work. In order to wake the corms in early spring and get them to grow, we needed a warm place. And Belgian endive has to be blanched—deprived of any light—to keep the leaves from developing chlorophyll, and becoming unpalatably bitter. A few weeks before we wanted to eat them, we built a sturdy lidded plywood coffin near the woodstove, filled it with fresh potting soil, planted the corms in it, in a grid, and kept them well watered. In a few weeks, they sprouted a white crown of tightly packed leaves that we harvested for our members and ate dipped in homemade sour cream with finely chopped leek and a sprinkle of salt. I could tolerate, barely, the mess and earthy stink of that soil coffin for the crisp taste of fresh vegetables during mud season.

The endive had taken the place of two barrels of sauerkraut that had spent part of the winter fermenting there, filling the whole house with the smell of old socks. Kraut was the practical counterpart to our froufrou endive project, an efficient and nutritious way to preserve cabbage, which we could grow in great abundance in our soil and climate. After frost, we'd cut the voluptuous heads from their wrappers of leaves and trim and clean them. Then they were ready to be shredded, salted, and packed for fermentation. The magic of traditional lactic-acid fermentation is that it not only preserves a fresh crop but transforms it into something entirely different in taste and even more nutritious than it was in its original state. Sauerkraut has more vitamin C than cabbage, more than almost any food on the planet, and contains a rich, living community of beneficial bacteria that are good for your gut. Its sour, salty, acid tang accompanied most of our meals.

We made sauerkraut for our members every year, but that year, we were at an awkward scale, well beyond the size of equipment made for large families and still too small to be properly commercial, even if we'd had the money for industrial tools. We improvised and temporarily turned the house into a wild sauerkraut factory. One day my friend Kristin Fiegl came over with her son, Jameson, who was seven. Ronnie came over too, bringing a pan of brownies and her favorite knife. Ronnie had been a schoolteacher until her retirement, and came over to help with the work and the children on days like this when there was a big job to do. We cleared all the furniture out of the dining room, cleaned it from top to bottom, and set up a station for chopping, one for shredding, and one for weighing and salting the shredded cabbage. Jameson was old enough to help, ferrying cabbage from the cutting board to the shredder. Jane played with Jet in the stray cabbage leaves that fell under the tables, and Miranda slept through the commotion in a bushel crate in the corner.

The hardest part of making kraut is packing the shredded, salted cabbage tightly enough into its container. You have to get rid of the air, and bruise and break the cabbage, in order to allow the salt to draw the liquid from the cells of the leaves. The brine must cover the kraut completely for it to properly ferment. If you are making a quart or a gallon of kraut, it's easy enough to crush and pack it with your hands and the end of a wooden spoon. We were making a hundred gallons. The packing was wearying. We tried using heavy wooden mallets, throwing them down hard into the barrel after each layer of cabbage was added, but it was exhausting and inefficient. Mark, who hates an inefficient system with the same passion a preacher feels for the devil, came in and found us sweating and panting with our mallets and decided there had to be a better way. That's how I found myself standing inside a fifty-gallon barrel in our cabbage-strewn dining room, wearing a miniskirt, with freshly scrubbed bare feet, stomping the cabbage down into its chilly brine like a deranged Germanic bacchant. It worked so perfectly, I am shocked there are no records of European villages holding annual kraut-stomping festivals.

We were near the top of the first barrel, my feet red and cold, when we heard a car pull in. Mark, who had been working the shredder, looked out the window and saw the seal of the state of New York on the side of an unfamiliar van in our driveway, and several men climbing out, clipboards in hand. I leaped out of the kraut barrel as Mark was opening the door to a whole parade of food inspectors from both the federal USDA and the State Department of Ag and Markets, on a surprise visit to check on our milk house, butcher shop, and storage facilities, and try to get to the bottom of what we were up to at Essex Farm.

They must have wondered why I was barefoot, dripping with brine, but I think they were so disoriented by the total weirdness of the scene they'd walked in on, and by the sheer number of potential violations all around them, that they couldn't bring themselves to ask. Jet stood in the middle of the pile of cabbage leaves, waving his tail diplomatically. I froze, wet strips of salty cabbage falling off my bare legs. Ronnie read the room and hustled the children outside. After a few stunned seconds, Mark started talking. I have no recollection of what he said. But he swept them out of the house very quickly, on a tidal wave of words. Then he walked them all over the farm, talking the whole time. He showed them the buildings, and our cobbled-together infrastructure, and described our business model.

Somehow, by the end of the day, we had the foundations of a good working relationship with all of our inspectors. We drove them crazy because we were always pushing the boundaries of what was allowed, and neither our business model nor the food we produced fit neatly into any of the boxes that it was their job to check. But they worked diligently with us to find solutions that would keep us in business. By the next year, we'd be large enough to rent a proper commercial kitchen to make our sauerkraut, but I think the year we stomped it by foot in the dining room made the best kraut of all.

Nathan took a week off to visit other horse-powered farms all over the Northeast. The day he was due back, Chad mentioned that the furnace had gone out at the Yellow House. Chad was on his way to visit family for a few days, Tim had a girlfriend he could stay with, and Blaine had moved in with Tobias above the post office. But

Nathan would return to find the house cold, and we should probably warn him. After work, Racey walked to the Yellow House and left a note on the kitchen table. "If you need a warm place to stay," it said, "you're welcome to come over." She went home, climbed into her ugliest pajamas, and went to bed. At nine-fifteen, there was a knock on her door. It was Nathan, carrying a bottle of her favorite Scotch. They stayed up late, talking and laughing. Then he said the words she had known were coming: "I like you and want to get to know you better and—" Racey interrupted him, talking fast, a blur: *Paratrooper. Work. Friends only. I can't!* He listened, then got up. "Okay, fine," he said. "I'll wait until you're ready." He turned to leave. "What? Wait!" she yelled. "Where are you going?" After that, they were together. Nobody needed to wonder, and they felt no need to define their relationship because it was obvious. They fit.

At Eastertime, Jane went to play at Ronnie's house with Ronnie's grandson, Jason, who was eight. They made cookies, and Ronnie sent a plate of them over to her next-door neighbor, Myrna, via Jane and Jason. Myrna took the cookies and asked Jane if the Easter Bunny was going to visit her house that weekend. The question confused Jane. Maybe we hadn't yet taught her that tradition, or else she was too young to remember the Easter basket she'd received the previous year. But she wasn't too young to remember the rabbits we had raised, nor how delicious they'd been. "No," she told Myrna cheerily, "I think we ate him." Myrna reported that Jason, who was a big fan of the Easter Bunny, was aghast. So we had left out some pieces of a typical American childhood and added a few unusual extras.

The farm was busy, the house was messy, the children were small and helpless, but we were doing okay. Miranda was sleeping in slightly longer stretches during the night. The brittle edges that come from prolonged fatigue were beginning to soften just a little. The snow disappeared, the mud dried up. It felt for a while like we were through the worst of it. When the peepers sang from the back pond in April, the fields were ready to work. And then it started to rain.

CHAPTER 8

At first it seemed like the rain would have to stop. And then like it would continue forever. Storm followed storm. Between storms, we had days of heavy drizzle. There was no hint of bright sun or warmth. The fields were saturated. The grass came along, slowly, but the animals stayed confined, in small paddocks or inside the barn, to keep them from churning the pastures into mud. After a huge storm passed, Mark and I loaded the girls into the car to drive around the neighborhood and gawk at water.

We got out at the bend in the Boquet River a

mile to the west and walked down the slick bank to the edge. In summer, this was our swimming hole, a calm pool we shared with the resident snapping turtle and a few hungry leeches, always cool at the bottom and warm at the top. Now the pool and its ring of rocks were gone. The river roared by in violent torrents that made me clutch at the back of Jane's coat. It was carrying whole trees that had been ripped from the flooded banks, and unaccountable tons of topsoil that turned the water the color of coffee, the spray a dirty white. Under the extraordinary static-like noise, we could hear a series of dull heavy thuds, the sound of boulders rolling along the bottom, cracking against one another, reshaping the river for the next thousand years.

We drove to the shore of the lake, a mile to our east, and stood next to it. There was no wind, and the water was a still, ominous swell. It had just broken over the high-water mark and was rising. "Remember this," Mark told the girls. "You'll never see anything like it again in your lives."

As the days and weeks ticked by, we began to feel like we were under a bad spell. It rained all through April and into May. This was planting season, and in a normal year, we would have been in the fields every moment of the long day and then some. Instead, we busied ourselves with indoor work and listened to the rain on the roof.

Lake Champlain continued to rise, surpassing the high-water mark by an astounding seven feet. Our neighbor Old Steve said he had never seen such a bad year for farming in all of his ninety years. "Forget farming," he growled. "It's not even a good year for getting your laundry dry."

Every morning as I made my coffee, I looked for the pair of robins building a nest in the lilac tree just outside the kitchen window. It was such a good nest. The outside was neatly woven from twigs and baling twine. The inside was lined with dog fur. The female laid three sky-blue eggs. After they hatched, I watched the mama bird carry worms and bugs to three open beaks. Would they fledge, or would they get worn down by the constant rain and die? I became inordinately attached to the outcome, as though those hatchlings were sages auguring the outcome of our year. Just as the window for planting corn closed without any corn in the ground, I saw our cat Penelope crouched in the rain with a robin chick in her mouth. The nest was empty. She must have eaten all three, one by one.

Still it rained. Houses along the lakeshore flooded. When the wind came up, waves crashed into people's living rooms. The restaurant at the town marina was underwater. Then the lake rose over the ferry dock. The ticket office with its neat lace curtains became an island, half-submerged, the dock itself three feet underwater. Mark went out on his sailboard and cruised over it, through the debris and floating logs. He looked down, saw the parking lot underneath, and enjoyed one brief moment of awe before his mast hit the power lines and he crashed.

On the farm, the rivulets that drained the sugarbush hill had turned to turgid streams. We ran out of hay to feed the animals, who were still confined. One night after the kids were asleep, I walked through the rain to the pasture we called Fallen Oak Field. It was a little higher than the others, and sloped very slightly to the south, so it was always the first one ready to graze in the spring. I was hoping against reason that it would be firm enough for the cows, who were consuming expensive purchased hay. I stepped gingerly into it. The grass was four inches high, but underneath, the soil itself seemed swollen. I felt the top layer of sod tear beneath my foot and slide off.

There's an ominous saying in our region, "A dry year will scare you, but a wet one will starve you." We don't face the same sort of years-long droughts that other regions of the country and the world do. The grass and the vegetables grow slowly in a dry year, and sometimes the vegetables develop intense bitter flavors; the pastures need to be carefully managed, the grazing animals moved off before they eat too low and damage the roots. Sometimes the hayfields don't regrow fast enough to make a second cut of hay. But the weeds grow slowly in a dry year, and the soil retains its natural structure. Roots grow deep into the still-loose ground, searching for water.

A wet year is a much bigger problem. A wet year brings funguses, blights, and putrefaction. Hay grows rank or is cut to rot on the ground. Tractors get stuck and rut up the field. Underground, there's no air for the roots to respire, so plants yellow and die. And the windows for tillage are tight, so it's tempting to work the soil when it's sodden, which compacts it. Compacted soil takes years to recover, through the patient work of roots, worms, and microbes. A wet year is hard on morale and also on our bodies. Our hands crack and get infected from being constantly wet and soft, and our boots carry an extra ten pounds of mud everywhere we go. It's hard to like farming in a wet year.

Eventually, we had to plant despite the conditions, or it would be too late. We did it gingerly, into mucky beds that we made behind the horses with the hiller. We couldn't plant field corn because it was too wet for the planter, but we planted an acre of sweet corn by hand and hoped. We watched the tomato seedlings turn yellow. Mark and I went to the field one evening after the kids were asleep to see how the seed potatoes were faring. They'd been planted on the highest ground we had in production, and I was hopeful. But when we dug into the ground and found one, it had rotted into a mushy, sproutless corpse that smelled like wet death. The whole acre was the same way. The smell of it got into my hands, and I couldn't seem to wash it off.

Strange things began to happen. A sow died for no apparent reason. She was eating, nursing well-grown piglets, carrying plenty of weight, looking fine, and then suddenly, she was dead. One of the new dairy cows got horned in her udder by one of our old dairy cows, and her milk turned the sick pink color of strawberry Nesquik. Then a pretty Jersey cow named Carmen, recently bought from a neighbor, gave birth to a little bull calf and immediately came down with diarrhea, stopped making milk, and got shockingly thin, even though she was eating like crazy. Dr. Goldwasser came and tested her for various diseases. The one that came back positive was Johne's. That was terrible news. Johne's is a mycobacterium, in the same family as leprosy and tuberculosis. Unlike those diseases, Johne's has no effective treatment. There's nothing to do once you know a cow has it except put her down. A cow can carry Johne's for years, dormant and yet contagious; once it breaks—which usually happens when a cow's immune system is stressed by a life event like calving or moving to a new farm—it's fatal.

Moreover, Johne's is extremely hard to get rid of, for a confounding fistful of reasons. Cattle are susceptible to infection only when young, but the disease isn't expressed, nor can it be detected, until they are at least two years old. At that age, most of them have already had their first calf, along with a good chance of passing the disease on to the calf through her colostrum, not to mention all the young stock through her manure. The best test available is maddeningly inaccurate, detecting the disease in positive cows only about half the time. Some cows become silent carriers, shedding huge amounts of the pathogen in their manure but never developing symptoms. Finally, the pathogen itself is so hard to kill that even lime or bleach won't manage it reliably. On pasture, it is known to survive through hot summers and cold winters for a year or more.

This isn't a rare disease. Some farms, once infected, just try to manage it and take the loss of production as a cost of doing business. Seventy percent of dairy herds in New York are thought to have at least some infected cows, but we didn't want to be one of them. Because of the way we farmed—milking fewer than two dozen cows, without antibiotics, on pasture, making raw milk for our members—we needed a herd that was healthy and strong, not compromised by the pressure of a chronic disease. So we culled Carmen, as well as her bull calf, and dropped them in the compost pile. Then we called the state veterinarian and Dr. Goldwasser. They helped us set up protocols to try to stop the disease from spreading. We began the expensive, heartbreaking process of trying to eradicate it. The next year, we'd have to cull a third of our herd because they tested positive. After that, it began to get easier, but it took us five years of hard work, diligent management, and ruthless culling to get rid of the disease.

Meanwhile, Chad's mare, Fern, was not well. Could it possibly be worms? Spring is the season when worms break their dormancy, and in a wet year, they can get out of control. As Chad, Racey, and Blaine worked in the butcher shop, I pulled out my microscope, graduated cylinder, and floatation solution. Then I put plastic bags and a Sharpie in my pocket, loaded Jane onto the back of the pony and Miranda into the backpack, and hiked to the field where the horses were pastured, to collect manure samples. These were my new jobs, a step down from the hardest work.

It had rained yet again the night before, and the fields were still too wet for the horses to work. The whole herd was grazing along a stretch of brushy pasture to the north of Pine Field. We knew we were sacrificing a few years of production in that section, as their hooves would churn the topsoil to mud and compact the whole area, but there wasn't a better option. At least the horses looked happy, tearing at the tender shoots. We turned the delighted pony loose to graze with them, and I sat with Miranda on my lap, chatting with Jane, while we waited for fresh dung. The three of us lay in the grass, faceup, watching the muted light play in the leaves of the ash trees above us.

The ash trees were not doing well, infested by a blight that had left the lower branches bare, riddled with rot. Worse was coming for them. An invasive insect called the emerald ash borer—which sounds innocuous, magical, even, like something from the Land of Oz—was beginning to devastate forests and the hedgerows of farms all through the Northeast. As climate change allowed the insect to survive farther north, the ash trees were dying. A downy woodpecker beat a joyful tattoo in the tree above us, needling his beak through the soft wood to reach the insect life inside. The tree's loss was his great gain.

Ash was the dominant species in our hedgerow. Ash trees grow tall and straight, good wood for tongues of horse tools, as well as decent stovewood that could be cut in the spring and seasoned, ready to burn after one summer, or even burned green in a pinch. We had warmed our house with freshly cut ash more than once, when we were caught short of wood at the end of winter. All these trees would be gone, the biologists predicted, in a decade. Such little things, these insects, would transform the landscape more radically than we could, with all of our iron and diesel. It messes with our illusion of control, the fundamental human belief that we must be masters of all we survey. Don't underestimate the power of the small.

Once I found a live worm in the toilet after one of the children had used it. At first, staring down at it, I tried to make myself believe it was something else, anything else. Or that it had arrived there somehow under its own wriggling power and not through my small child's digestive system. Despite my own highly developed powers of denial, it was a worm, a long one, unmistakable. And there was no other possible explanation for its presence. I was filled with horror and shame. *This is what happens when you're not like other Americans,* I thought. *This is what you get for raising children on a farm.* It wasn't the first time that sort of thought had occurred to me. So many of the choices we made—the very choices that added beauty and value and satisfaction to my life—seemed to go against what the prevailing culture told us was right. Who said you can leave the food system, grow it instead of buying it the way it's meant to be bought, in neat plastic packages? Who gave you permission to give this child raw milk that you coaxed, yourself, from a beast's hairy udder? A part of me—the rule-bound, middle-class good citizen—was always awaiting punishment.

Through my mind flashed all the times I'd let the kids kiss a newly hatched chick or walk barefoot in the pasture or failed to make them scrub their hands well enough before dinner. I thought of them as babies, sitting between the rows of strawberries in June, pushing handfuls of soil into their mouths along with all the berries within their reach, so that their faces were streaked in red juice and mud, and how, because they were so determined and I was so busy, I'd let them do it.

I ran to the barn for a pair of shoulder-length gloves, then fished out the worm and put it in a jar. All my anxieties about values and normalcy and parenthood were contained for a few hours in that cylinder.

Mark never felt these things. His own upbringing had been, in most ways, outside the lines of normalcy. His parents were back-to-the-landers who had left city careers for a piece of shaley land in the Catskills before Mark and his sister were born. They raised much of their own food and lived, for the first part of Mark's childhood, in a converted barn without indoor plumbing. Also, Mark's nature was to question all the rules, all the time. He was naturally distrustful not of the hairy udder but of the neat plastic packaging.

I dialed our pediatrician, Dr. Beguin, feeling

relief in the idea that he knew us. That seemed important, in case this was the sort of thing that had to be reported to the authorities, the ones who can take your kids away. When his nurse, Carolyn, answered the phone, I explained what I had found. "Hm, impressive, aren't they?" she said casually. "Why don't you bring her in this afternoon?"

Dr. Beguin, in his gentle monotone, talked me through it, all the facts I knew when dealing with animals but, in my panic and anxiety about parenting, had failed to apply to my own child. Parasites aren't entirely evil. We have evolved together, we and our parasites, and as long as we stay in balance, there is no cause for concern. In fact, there is evidence that exposure to parasites can be good for us, because they help train our immune system to focus on the real pathogens and ignore the rest, which reduces the overreaction that causes allergies and autoimmune disease. Maybe that was why the children had been so instinctively compelled to push soil into their mouths in the first place. In any case, these things were unrelated. The worm I had found was a roundworm. It was far more likely that she had picked it up from the sandbox at the playground than on the farm. It did not make me a bad mother. I could be grateful to the worm for doing its job and move on.

• • •

The horses grazed around us, making greedy ripping sounds with their teeth. They were eating young fescue, mixed with reed canary. It was plentiful, and in the sweet stage of growth that drove the horses crazy with pleasure. They would turn their noses up at the reed canary in a few weeks, when it became tall and coarse, but for now it was a delicacy. Jake lifted his tail and contributed a manure sample, and then Belle, the pony, and finally, near lunchtime, Fern, my target. I gathered balls of dung the size of shooter marbles from each pile, labeled them with the horse's name, stuck them all in the back of Jane's backpack, caught the pony, and headed back to the house.

There is a small scientist inside of me, a scrap of a younger self who could have chosen that path at the fork in adolescence and been quite happy. I like the equipment, the methods, the rules, the moments of quiet concentration that lead to the discovery of useful information. I like questions that have sure answers. After lunch, with both children napping, I spread out my supplies. I wasn't looking for worms but for their eggs. A few worms produce a few eggs; a lot of worms produce a lot of eggs. A heavy worm burden can make an animal look worn down, but you can't tell for sure how infested an individual is without looking and counting.

I weighed out a gram of manure on a digital scale. Parasites are a nuanced subject. Every grazing animal has some parasites. The goal is not to eliminate them but to manage them so that the worms are not a drag on the animal's overall health. There are enormous variations in an individual's ability to resist parasites and maintain good health. Youth is a dangerous time, as are pregnancy and parturition, when the immune system lowers its defenses to allow that greatest of all parasites, the fetus, to thrive. Malnourished animals are vulnerable too, because parasites tax the blood supply, syphoning protein. If protein is already in short supply, you have a situation on your hands. Young animals will not grow. Older animals will lag, look poor, sometimes die. Vulnerability to parasites varies from species to species. Cattle are hardy and rarely need treatment. Sheep are another story. Worms are the bane of the shepherd in our climate, where sheep are grazed in small lush fields instead of over expanses of arid rangeland. And horses, even giants like ours, can be brought low by worms. Sometimes it is sudden, dramatic, and fatal. The collective power of the botfly larvae can block digestion; a knot of strongyles can explode the aorta. Do not underestimate the power of the small.

I used a dropper to put some of my manure slurry on a slide and waited for a few minutes for the eggs to float to the top, then placed the slide under my microscope. The slide holds one milliliter of solution and has a grid printed on the top, which makes the job of counting eggs easier.

Before modern worming medications, farmers used all sorts of mixtures to kill worms. Tobacco was the old standby—plug tobacco was readily available, cheap, and effective. The problem was that there was a fine line between a purgative and a poison. Now there are anthelmintics, with a wide range of safety and a high rate of efficacy, but we have a different problem on our hands: resistance. If we dose every animal in the herd whether they need it or not, we selectively breed worms that are resistant to wormers. In many places around the world, whole classes of modern anthelmintics are now entirely useless. But if you dose only the few animals whose parasite burden is too high for good health, the rest of the herd will still harbor naive worms, so the evolution toward resistance slows down. It's not as easy as dosing the whole herd—it requires judgment, and close observation, and looking at the big picture over short-term gain—but it takes into account that the genius of nature is its mutability.

We always want a simple answer, but what we get, over and over again in food production, is complexity. Is that a strongyle egg I see on the slide, or is it a piece of pollen? Is this horse infested enough to treat? With what, and what else can I do to mitigate the problem?

The white pony's slide was clean. Fern's, though, was a bit concerning. There were strongyle eggs all over the place. I told Chad what I'd seen, and he dosed her with fenbendazole, moved her to fresh pasture. When we tested her again two weeks later, her egg count was down to near zero, and her coat was glossy.

In late May, we got a break. The sun came out, and the wind came up and dried the fields. The water in the lake continued to rise for a while, as the streams drained the valleys, and then began to subside. Suddenly, there was action everywhere. The work we should have been doing for the past eight weeks needed to be done now, all at once. We hitched Jay and Jack to a cultivator fitted with inward-facing discs, to shape raised beds in tiny Home Field, the only place dry enough to consider putting plants, and transplanted the overgrown lettuce and onions from the greenhouse. Tim hitched his young team of gray Percherons to the manure spreader, and spread loads of compost on the driest parts of Monument Field, where we expected to grow most of our vegetables, pushing through the work until the wheels and the horses' hooves sank too deep. We threw a lot of resources at the acre of sweet corn, which had miraculously germinated, stoking it with compost, beating back the weeds, then adding fake snakes made of old hose and strands of flashy Mylar, to scare off the crows. Chad put Jake and Abby on the forecart and dragged the pasture in Essex Field, to knock apart the cattle's winter manure, then added his gelding, Arch, to the hitch, three horses abreast, and pulled the big spring-tine harrow over the headlands and the asparagus patch, trying to gain control of the weeds, which already had the jump on us.

All parts of the farm were interconnected. The repercussions of the rains would be myriad and would affect us all year. The budget had taken a hit already because of the hay we'd bought to get the animals through until the pastures were dry. The greenhouse flats were all full, and so was the greenhouse itself, and we couldn't transplant things out, so we couldn't seed anything new, which would push the timing back for everything.

The dry weather held for a week. The dairy cows finally went out to pasture, galloping and bucking awkwardly at their first sight of it. We transplanted and seeded maniacally into the cold wet ground, attacked the weeds with the desperate energy of a village about to be overrun by invaders. Everyone started in the dark and worked absurdly long days, ate their meals in the field. The taste of that spring in my memory is cold buttered biscuits a little gritty with soil, eaten over windburned lips; or slapped-together lettuce and mayonnaise sandwiches with a hint of leather and horse sweat. The children came to the field with us every day, barefoot. Jane tagged along with me or Racey, and we plopped Miranda down in the field near whoever's work was most stationary. She crawled along the muddy furrows, pushed herself up on her feet to test her legs for walking, ate alarmingly large fistfuls of dirt. At the end of the day, I stripped both girls at the door and put them directly into the bath to soak off the mud.

Mark ran from job to job, hustling, smiling, singing to himself. That first planting of potatoes had rotted in the ground, and it was getting late for potato planting, but they were an important crop to get us through winter, so he hitched Jake and Abby to the potato planter and put in a Hail Mary patch. The horses had been working as hard as we had been, and all their youthful freshness had been buffed off by it. When he told them to whoa, they did it right away and stood perfectly still, breathing. Mark was working alone, so he thought he could trust them to stand in the field and rest while he ran to the headlands for another bag of seed potatoes to refill the hopper. It was a Friday, and our members had come to the farm to pick up their food. I was in the pavilion, cutting more seed potatoes and helping people with their food, when I heard a metallic cacophony on the road in front of the farm, accompanied by the terrifying sound of eight hooves at a dead run. I looked up to see the horses with the potato planter clanking along behind them, Mark far behind, at a sprint. I had a vision of them turning down the driveway toward us, our members and their children in their path. Instead, they ran past the driveway and crossed over the opposite lane and onto our neighbor's lawn. The left line tightened on something, and they were running counterclockwise circles in the turf that grew smaller until the planter hit a stump and stopped. Mark and I reached them at the same time. They were blowing hard but unhurt. Mark checked the equipment. Except for a bent bar, the planter was somehow, miraculously, intact, so he went back to the field and finished planting.

It turned out that Mark's energy was not bottomless after all. He burned with such brightness that spring, but eventually, the fuse blew, and the whole room went dark. He spent a long day with his six-feet-six-inch frame perched atop the bumpy two-horse cultivator, trying to knock back quack grass; it's no use getting seeds in the ground if they're just going to be vanquished by weeds. Then he jumped off the cultivator to plant next year's strawberries, which had been languishing in the cooler since they'd arrived from our supplier and were beginning to look limp. He was using a spade, making holes in the wet ground, pressing the plants into them with his heel. Somewhere around the four hundredth plant, his back twanged, and the shock of it sent him to the ground. He lay there, trying to relax his spasming muscles. The to-do list was still long, and rain was coming again, so he got up and pushed through the pain, planting until well after dark. By the next morning, he could barely move. I left him with the kids while I went to milk the cows. When Miranda woke up crying, he crawled into her room, tipped her crib over, and gently rolled her out because he couldn't stand up to lift her.

I was hoping it was just a three-day tweak. That's what we call it when a horse pulls or strains something and needs a little rest. In the meantime, Mark could run the farm like a brain in a jar, sorting out the priorities from the office window, directing without doing; we had six other fit people to do the physical work, and four teams of strong horses. In any case, it started to rain again. There wasn't much we could do.

I wish I could say that I was a supportive and empathetic partner while Mark was injured, but seeing him in pain brought out the worst in me. It cracked the image I'd constructed of him: so tall and strong and determined and capable of using his body to wrestle to the ground whatever problem the farm threw at him. I had seen him do extraordinary things over the years—move rocks all day without resting, hoist an enormous beam onto his shoulder, work sixteen-hour days for weeks—and had trusted it would always be so. Now that superman was replaced by this immobile and needy *imposter*. I couldn't reconcile those two different men, and it frightened me, and my fear made me brittle. We began to argue about small things—what the children were eating for lunch or if they could watch a video while I milked—and the fighting made him tense, his back worse.

Two nights later, I was yanked out of sleep by a disconcerting sound. There were only two sounds that could wake me that fast. One was a child's cry. I'd begun to think there must be an undiscovered organ, adjacent to the uterus, that is tuned to the sound of a child's need. The smallest cry, or a feverish whimper, would yank me out of bed and have me on my feet before I was fully conscious. Other sounds—the howling of coyotes or the sound of Jet chasing them out of the barnyard—would register but leave me where I wanted to be, near the bottom of the ocean of sleep.

That night, it wasn't a child but the other noise: the dreaded sound of sharp teeth making expert work of a piece of wood disconcertingly nearby. I had developed a good ear for rodents. I could tell a mouse's covert nibble from a squirrel's hyper scrabble while only half-conscious. Either of those sounds would allow me to sink back to sleep—a problem that could wait until morning. But rats were another story. This was the unabashed racket of a diligent worker at a construction site, with a permit. There was nothing covert about it. Rats sound *entitled.* And that was the noise pulling me to the surface of full consciousness that night and holding me there.

All farms have rats. We do our best against them, and good farmers keep them under control, but inherently, farms are rat utopia, with food available year-round and soft warm places to bed and mate. When we first moved to our farm, there was a legacy population of lazy rats who lived happily in the old granary. They had reproduced for many generations without a challenge, and we vanquished them fairly easily with traps and cats. But the survivors regrouped, rebred, honed their genetics to our challenges. Whenever we let our defenses down, the rats surged. We'd learned that a sloppy grain delivery, an unturned pile of cool compost, or the inattention of a busy spring could cause a population spike.

The previous fall, we'd decided to buy our greenhouse supplies in bulk, when the price was cheapest. We'd bought twenty yards of potting soil, delivered in a dump truck and placed between two giant tarps spread on the ground next to the greenhouse, across the driveway from our house. That pile of potting soil was the best rat home ever engineered. It was rich and loose, and the top tarp kept the soil warm. The chickens were wintering in the greenhouse, their feed stored nearby, so there was an all-you-can-eat buffet mere steps away. At the end of February, when Mark pulled the tarp back to begin seeding the early crops in the greenhouse, he discovered that they'd spent the winter having orgies, holding feasts, sculpting intricate living quarters. He and I had stormed them one bloody afternoon and dispatched the majority with sticks, shovels, and Jet, but a few had escaped and dispersed.

The noise I heard was likely made by one of those exiled survivors, one that had discovered that our house was yet warmer and drier than their old heap of soil, and had decided to get the place ready for his fellows. It sounded like a major excavation of the laundry room floor, the widening of a hole. I nudged Mark, who reluctantly heaved himself from bed and limped down the hall. I followed, tentative, miserable. Outside, I could handle them. Inside, rats were different. They made me want to cover my face and jump on top of furniture.

Mark was in no shape for a round of hand-to-hand combat. He could barely move. He shuffled into the laundry room, looking under items near the source of the noise. When the rat squirted between his legs and into the bathroom, I screamed. He hobbled in front of me and shut the bathroom door. "Go in after it," he said. *That,* I thought, *I cannot do*. He was injured, and I knew I *should* go, but I simply couldn't make myself.

So he did it. I heard a cartoonlike scuffle behind the closed door that went on for ages. "I need some help in here!" he yelled. I paced and pretended not to hear. Finally, he emerged, the toilet plunger in one hand, a huge dead rat, held by the tail, in the other. The rat had scurried underneath the linen chest, and Mark had shoved it aside, whereupon the rat had scrambled directly up his leg and leaped into the bathtub, which was the rat's biggest strategic error, because the slippery sides contained it. Mark grabbed the plunger, and that was the end of the rat. But Mark was definitely a casualty. The battle had thrown his back into a new range of pain. The next morning, for the first time since I had known him, he did not get out of bed.

Mark wasn't used to pain. His body had always done for him all the unreasonable things he'd asked of it. Suddenly, it seemed, his body had decided that the generosity of its youth was spent. He couldn't stand up and he couldn't lie still. He moved his legs constantly, looking for some relief, until he wore a hole through the sheets.

The pain started in his body and quickly bled to his spirit. He stayed in bed most of the day, shades down, not sleeping. He played endless games of electronic Go on my phone. His muscles began to waste. My tall oak of a husband looked weak and undernourished. The pain took away the center of him, so I couldn't find him in his eyes.

And he didn't seem to care what happened on the farm. The rain eased up, and we were back to work in the fields without him. I'd ferry bunches of questions from the barn to his bedside. "I don't know," he'd say. "You'll have to figure it out." This was a man known for managing all of our many endeavors down to the microscopic level. His usual optimism evaporated, along with his leadership. Farming was the singular passion that had animated him since we'd met. Without it, I hardly knew him. Not that I had time for contemplation. I had a baby, a three-year-old, and Mark to care for. And the only remnant of his old self I could recognize was his formidable libido. When I walked too close to his bed, carrying a dirty diaper, stupid with exhaustion, he would reach for me hungrily. I felt like I was stuck inside a three-part fugue of need. The soprano sang, "Nurse!" the alto went, "Read me a story," and the bass boomed, "Want to make out?" all at once, over and over and over again.

It wasn't like we could just call the year a loss, sit it out, and take a do-over. We had a mortgage to pay and a lot of people to feed. Some of them had paid us the whole year in advance so we could meet our spring expenses. They trusted us to put food on their tables.

There was another thing, and it was so scary, I couldn't look it square in the face. We'd gotten a tax bill that we weren't expecting—on the grant money we'd received for the solar panels. The expenses had come in one calendar year, the reimbursement in the next, and that made it look like we belonged in the same tax bracket as a mid-career banker. When I saw the number, I had to sit down. We didn't have the money to cover it, not even close. I put the bill back in its envelope and stuck it behind the flour canister in the pantry, where I wouldn't see it, except in the bad dreams that woke me in the defenseless hours between midnight and dawn.

We met without him in the mornings to sort out the work, the six of us around a big pot of coffee. Managing people was not a strong part of my skill set. I hated telling the farmers what to do or trying to defend decisions that I wasn't entirely sure were right. And though I didn't like to admit it, I wasn't good at the complex organizational work of keeping the whole farm going, balancing all the competing priorities for the good of the whole, especially in such an anomalous year. Mark was a genius at that. He had the ability to see and quickly rank the vast number of jobs that needed to be done. Moreover, he'd never had a problem with telling people exactly what to do. Occasionally, he was wrong, but he was never in doubt. I had spent most of my time working with the animals. The technical aspects of plants—the timing of planting, the spacing between rows, the temperature that a seed needed for germination, the quantity we needed to plant—those things were a mystery to me. I felt entirely lost and very much alone. I should have summoned more empathy for his pain, but what I felt instead was smoldering resentment.

When we had a dry day, we needed to get as much land prepared for planting as possible. I was very lucky to have such a dedicated and capable team of farmers with me. They used all their resources to help keep the farm going, and worked so hard and thoughtfully. It wasn't their farm, but they acted like it was. One rare dry day, Chad figured out how to hitch six horses to a tractor-size spring-tine harrow so he could more efficiently knock back weeds and loosen the soil in Monument Field. Blaine, Tim, and Nathan followed, making beds raised up out of the moisture, then cultivating them with another team of horses, to smooth them for planting. They got thirty-one beds prepared in one long day. Tobias and a team of volunteers did the hard handwork, hoeing rows that were too wet or too far gone with weeds to be cultivated with the horses. Racey moved fencing so we could get the beef herd out of the barn and onto the grass at last. And grass, at least on the better-drained fields, was the one piece of good news. It was lush and high, and the dairy cows were gorging on it, producing lots of delicious milk.

The farm was a large ship that had lost its rudder. No matter how hard we worked, we couldn't keep up. I was not getting nearly enough sleep. Weeds were encroaching, crops were drowning, employees were working unsustainable hours. In the evening, when the children were asleep, Mark would emerge from his bedroom and come downstairs to sit on a chair while I did the dishes. Our conversations were loaded with fear and shaded by fatigue and often ended in squabbling. I looked and looked and couldn't see a good ending. In bed, in the dark, I'd wonder what would happen if we left, the girls and I. What if we just turned our backs on all of this wet mess and drove away to find a new place and start over.

But I had chosen Mark and a farm life after having a good look around the world. I'd known I was trading the possibility of a nice steady paycheck, of weekends off and paid vacations, for work that was beautiful to me. I'd known that choosing farming meant choosing a modest life and sometimes a pinching and scrimping one. I had even known that this dark part of Mark lurked somewhere under the surface, the foil for his great passion, energy, optimism, and belief. I'd known all of it when I had made my vows to him in the loft of the West Barn in front of all those people. As the baby howled and the rain fell, I looked at the cards on the table and thought, *Sometimes the hardest hand to play is the one you dealt yourself.*

Twice in that time of crisis, our marriage was saved by women I love. Once it was my friend Cydni, whom I've known since we were eighteen, when we were freshmen-year roommates. She had returned to her small ranching town in Idaho right after graduation to marry her cowboy lover, so in the marriage department, she had ten years of experience on me, but she'd always been wiser even when we were young. The week Mark and I got married, she took me aside. "Listen," she said, "the first year is hard, and then it gets easier. Around the seventh year it gets hellish, and then it gets easier. Don't use sex toys until year eight. You have to hold something in abeyance." She is an attorney, and words like "abeyance" sound natural coming from her, even when she is talking about sex toys. This time, when I called her—from the only private place in the house, inside my closet—she could tell it was serious. "You have no idea how hard he is to live with,"

I whispered into the phone. “You aren’t so easy yourself, my dear,” she answered.

Another day, I called my sister. She knew all about Mark’s usual overpowering exuberance, and she listened to me describe this new person I was living with, his opposite. “He is so extreme!” I complained. “Yeah. You would never be happy with a normal person,” she said, sighing. The truth of both women’s words settled into the angry place in my heart and made me laugh.

Mark’s pain didn’t go away all at once, but it abated a little at a time. He saw a good physical therapist who gave him exercises to do and assured him he would heal eventually. The therapist said it was important to get up and move around even if it hurt. So Mark emerged from his bed, snappish and testy. The fields looked terrible, despite all our effort. We’d lost the second planting of potatoes to rot. The first planting of tomatoes had drowned, and so had all of the peas, half the winter wheat. Weed control was a nightmare. Cultivation works best in dry conditions. A lot of the weeds that we had disrupted with hoes or the horses had simply replanted themselves in the wet earth and grown right back. All the rain had washed nitrogen and other nutrients out of the soil, leaving plants weak and vulnerable to pests and diseases. There was downy mildew in the cucumbers, and battalions of beetles eating the leaves of the zucchini and melons down to sad lace. And we'd compacted the soil, in spite of our best efforts, which made it hard for the roots to penetrate. The crops had never looked so bad, yellow, stunted, drooping.

Mark looked at them and his shoulders slumped. The sight of the farm was yet another kind of pain. He cast about for someone or something to blame, and everyone around him received a dose. After how hard we had worked in his absence, that did terrible things to morale. Chad was quietly exhausted. He'd moved out of the Yellow House and was living in a barn with no kitchen, no bathroom, little money. One of our members came by in tears to tell Mark that Blaine had shamed her in the butcher shop for taking too much meat. Customer service wasn't Blaine's strong suit, with her fiery, volatile moods, and Mark squabbled with her about it, both of them dug into their own corner. But we didn't have anyone else who could cut meat. I braced myself against all the conflict, tried to hold it away from me, keep my eyes on the fields and what was happening in them.

Tim took me aside one afternoon while Mark was at physical therapy. We sat under the sour cherry tree in the scraggly grass in front of the house. "Mark . . ." he said. "Maybe he's just not that good with people. Be careful of getting too big." It was like an oracle was speaking hard words out of Tim's kind and handsome face. Were we growing beyond the bounds of what we were supposed to be? All the old arguments about scale played through my head. The thing about growth is that it's awfully hard to reverse. You can't service the cost of having gotten bigger by becoming smaller again. I didn't know how to answer Tim, but I worried that he would be quitting any minute, and the rest of the farmers would too, and Mark would go back to bed, and I'd be crushed by the weight of the whole place on my shoulders. I'd heard rumors that Nathan and Racey would be leaving that fall to work on a different farm down the road. I'd walked in on a quickly hushed conversation between Tobias and Blaine on the same subject. They all wanted their own farms and felt ready to move on. I understood, but I still felt betrayed.

One afternoon in June, the sky northwest of the barns turned the color of old bruises—dark shades of yellow, green, purple. The clouds moved fast, full of thunder, and when they opened up, they dropped hail the size of marbles. The thrum of ice against the ground in June may be the ugliest sound in a farmer's world. If hail destroyed the crops, it would be too late to start over. Mark had told me stories about fields he'd seen after a hailstorm in the Midwest that looked like they had been ravaged by a plague of locusts. We had just transplanted the rest of the tomatoes and eight thousand leeks.

After the storm ended, with the girls in bed, Mark and I went to check on the crops and to see if the broiler chicken coops were where we'd left them. On the way out, we stopped to look at a fungus we'd never seen before, growing on a composting pile of hardwood chips. It formed sloppy neon-yellow splatters in the wood. The broilers had survived the storm inside their mobile coops, and the leeks in the field were beaten but not dead. The tomato plants were all lying flat in the wet soil, but if we could get them trellised quickly, they'd have a chance of surviving.

As we made our way along the farm road, down the hill toward Fallen Oak Field, something odd caught my eye from fifty yards away in Long Pasture. Squinting, I saw one of the pretty little Jersey heifers on her side, dead, legs sticking out stiffly from the four corners of a grotesquely rounded body, as if someone had inflated her with a bicycle pump. I jumped the fence, speculating. Bloat? That pasture wasn't rich enough to cause a heifer to bloat. Predation? She looked turgidly intact. Then I saw the exploded tree, hunks of bark littering the grass. Lightning from a storm that had come through just before dawn the previous morning. I'd heard the rip and crash of the strike from bed. She must have been under the tree when it hit, but the force of it had thrown her through the air twenty feet. "Mother Nature," Mark said, "is pissed."

CHAPTER 9

The summer solstice came. Mark was in and out of bed, doing his exercises, worn down from the grinding pain. At the end of the year's longest day, we took a slow walk around the fields together. The plants were stunted little things. Some rows were completely empty because they had not been planted. "If it were mid-April," he said grimly, "I'd say it looks fantastic." How tightly our egos were tied to the fields. When the farm was thriving, it was easy to be proud of what we did. When it wasn't, we felt bad not just about the crops or sick animals, predation or weeds, but about *ourselves*.

When the rain stopped, it stopped so completely, the memory of it was like a bad dream. The wind came up, the sun pounded down, the puddles disappeared, and the topsoil dried and cracked. In place of the rain came oppressive heat and more weeds. Also the pressure to do the impossible: catch up with work that could no longer be done. There was no field corn in the ground; it was too late to plant. The acre of sweet corn was struggling. The grass in the hayfields was already stemmy and heading toward seed, so the hay we'd eventually make from it would be coarse and bulky, low in protein and available energy. But the weeds were flourishing. All hands were on deck to kill them as fast as possible. Mark's injury had reversed our roles: he spent long days in the house with the kids, and I spent them in the field.

It was a relief to be in the field, away from the house and the barnyard. The tension was thick in both those places. Mark and Blaine had a series of skirmishes over priorities, division of labor, pay scale, hours, and anything else they could manage to argue about. These fights would blow up and then blow over, but the sting of them lingered. Blaine and Tobias were already half-gone, I knew. I heard them in the office, making phone calls to Realtors, looking to buy a piece of land. But I worried that fighting with Mark would push them out early, and then what would I do? I didn't have the skills to cut meat. Tobias was managing the vegetables. There was no way I could handle that too. And the day before, I had heard Mark arguing with Racey in the front yard. She was in charge of animals that summer and had been trying to castrate a litter of piglets. The sow had escaped from the place where she was penned, and Racey had to scramble to avoid being attacked. She was asking Mark for advice, but all she was getting was criticism. She hadn't thought it through, he was saying, and had wasted half the afternoon, and we couldn't afford it in a year like this. There wasn't any concern in his words for the fact that she had nearly been savaged. I listened from the house, torn between wanting to hide and wanting to intervene. I took the middle ground, shouting out the window to them both to come in for a cup of tea, whereupon Mark hustled upstairs back to bed, and Racey and I talked through a better pig plan.

The day after the solstice, I got up at four and went out to catch Jay and Jack. Back before farming, when I was a civilian, the arrival of summer was a cue to take it down a notch, enjoy half-Fridays in the office, rent a beach house, relax. Now summer was the high season of work, the climax of the light.

Jay and Jack were in their twenties then, a pair

of old campaigners, their shoulders scarred by years in harness. The scars came from galls, and white hair sprouted from them that looked like medals of honor on the brown fur. Most working draft horses carry them. The old-timers would wash their horses' shoulders with salt water to try to toughen the skin under the collar in the spring. Everyone tries to prevent galls—with well-fitting collars and clean pads, thorough brushing before harnessing, and making sure there aren't pieces of mane stuck under the collar, which rub like a strand of rope in a sock—but sometimes they happen anyway, especially in the spring, when the hours in harness are long and the horses are soft and not yet in good muscle. Galls were hard to heal because the horses were working every day, and the scab that formed overnight would come off under the sweaty collar the next day. We dabbed salve on the trouble spots—a thick, ancient-smelling paste that came in a can whose design looked like it had not changed in a hundred years—but what helped most was several days of rest, and those were in short supply.

My relationship with horses had changed a lot in the years we'd been farming with them. I had stopped ascribing humanlike feelings and motivations to them, and now saw them fully for the beasts they were, which only made them more interesting. Working hard with them every day, I'd lost the reverence I used to feel around them.

But it was replaced with something stronger and more intimate.

Jay and Jack were still healthy and useful, but there was no joy in asking them for a full day of heavy pulling. They were very good for light jobs and for training new teamsters to drive. They were square-bodied, straight-shouldered, and so well matched that I had to look for the irregular whorl in the white hair on Jay's forehead to tell them apart. If I couldn't see the front of their heads, I could tell them apart by coming close to them and speaking. Jay was an amiable guy and would turn toward the attention. Jack, however, was a curmudgeon. In his stall, he flinched when I patted him, as though I were a fly, and gave me a look that conveyed not fear but very deep displeasure. He carried a narrow six-inch scar of bright white hair across the hard bones of his face, between the eye and the muzzle, and I had long wondered where that had come from and what it had to do with his disposition. We had bought the team from a retired couple in Vermont who had used them recreationally. They had bought them in Ohio from an Amish dealer. Everything we knew about their past came thirdhand. They were supposed to be Belgian/Suffolk crosses, and they had a Suffolk's chunky, short-legged, useful build. They had long, natural tails, which was unusual but a good thing. Most draft-horse breeders cut away some of the tail and keep the hair bobbed short, which keeps the long hairs from getting tangled in harness or machinery, but takes away his only weapon against the flies.

I left the house and walked toward the horse pasture in the half-light, pausing at the clothesline next to the kids' swing set to hang a load of sheets. Barbara was milking that morning, and I could see her coming up the hill toward the West Barn, the cows stretched out single-file in a long line in front of her. Barbara was our longest-term employee, our friend and neighbor. She was in her sixties, with two adult daughters. She'd grown up in Switzerland, then joined the foreign service and lived in Afghanistan, Iran, Poland, and Singapore before marrying a farmer who lived in the north country. They farmed together, three miles down the road from us, for sixteen years, until a drought finished the farm and their marriage. She moved to town after that, fledged her daughters, and then started working with us. It was good, she told us, to be able to do the farmwork she loved without the stress of endless decisions.

As Barbara and the cows passed the East Barn, there was a commotion. I could hear her yelling but could not make out the words. I ran toward them. The hens were fenced in a shady patch that included the compost pile, where they spent their days scratching. The nest boxes and night perches were tucked into the foundation of the old round barn, where it was dark, cool, and silent. There was a fox dragging a limp hen across the barnyard. He saw me, dropped the hen, and lit out for the high grass at the base of the woods. Barbara and I watched his red back crest the grass like a dolphin.

We returned to work, she with the cows and I with a set of wrenches, setting up the shields and sweeps on the horse-drawn cultivator for the day. When I was almost finished, the hens raised an alarm. The fox was back. I saw him scoot under the electric fence. The ground was dry and the fence charge low, and the fox's hunger overrode the shock it gave him. I ran toward him, grabbing a stick on the way.

We did our best to live in harmony with predators. Fox, coyote, raccoon, skunk. Sometimes bobcat, bear, or mink. The aerial hunters, owl and hawk. They were all a healthy and necessary part of the farm and of the larger ecosystem. They kept the population of rodents in check and the deer at bay. But any of them would take an easy meal if they could get it, and everyone likes chicken. To keep the hens safe, we pastured them inside electric fences, away from natural cover, and rotated their positions so the predators couldn't get entirely comfortable with a situation. Later, we added livestock guardian dogs. Usually, these deterrents worked, except when a bold or desperate individual broke the rules.

I jumped the fence and peeked into the hen-house, and there was the fox, his long body tucked into a nest box, his delicate face in poker mode, eyes flat, revealing nothing. Nobody here but us hens, it said. I menaced him with my stick as the hens and the rooster squawked their alarm. If Mark had been with me, he would have grabbed the stick, and that would have been the end of the fox. But I hesitated. What if he was rabid? What if he bit? What if I don't want to hit something so wild and scared? The fox read me, gathered himself. He sprang from the nest box, leaped past me, and was gone. In the moment he passed me, I saw that he was an awfully poor fox, young and small and stringy, his fur patched with itchy-looking mange. Though I knew better, I hoped he was an impressionable thing and would be scared of us and of the barnyard after this and leave the hens alone. I set the electric fence back into place, narrowed the gaps at the base, and made sure it was hot, then walked out to catch the horses.

I carried a reel of portable fence in one hand and four fence posts in the other, the horses' halters and lead ropes looped over my shoulder. The fencing was a concession to Jack's only flaw. We had owned him for six years, and I had wasted many frustrating hours trying to catch him in the field. This was a fault he had come with, and we had not been able to break him of it. If he was working every day, he was easy to catch, but if he hadn't worked for a week, he was almost impossible. He would wait for me to get close, then trot to the opposite side of the pasture and resume grazing.

On a day like this one, I didn't have the time to match wits with him. So I walked him into a corner of the paddock, pushing him in front of me, and then, using my reel of portable fence and the posts, made a triangle with Jack in the middle. It was small enough that he could not escape. He looked mildly peeved at this, the oldest trick in our book. I touched him lightly on his back and tossed the lead rope over his neck. Instantly, as always, he relented. That was the thing about Jack. As soon as he felt the rope over his neck—which was no physical restraint at all, only a symbol of it—he was caught, and he was all business. Aside from his rebellion in the pasture, he had perfect manners. He never balked or misbehaved. He was the first horse in the hitch to listen and do what was wanted of him. He loved rules, straight lines, and doing his duty. As a teamster, I had to be careful not to let him get ahead of me, because he was sharp-brained, would compare past to future and look for patterns in order to anticipate what was coming next. If Jack were a human, I'd hire him immediately and pay him well. He would be a dour presence but terribly efficient.

It was forecast to be the hottest day of the year, reaching close to a hundred degrees before noon—perfect weather for killing weeds. I harnessed as quickly as I could and walked the horses out of the barn. Then I hopped up on the seat, and we set out for the field.

The horses and I were dressed for the conditions. I wore a long-sleeve white button-down shirt with the collar turned up, and a giant hat to keep the dazzling sun off my head. The horses wore their old-fashioned fly nets. As I walked them out of the barnyard, I saw the fox again, crouched near the long coarse grass at the edge of the pasture, gazing chickenward. He was twenty yards from me, in full view, but he didn't flinch when I looked at him, didn't move when I yelled at him. Either hunger had made him very bold or he was ill. I couldn't do anything more with the horses on my hands, so I went on, wishing the hens good luck.

As we walked the half-mile to the field, I whoaed twice to slap the flies away from the horses' bellies. Even at that early hour, the sweat wore a groove through the dust on Jack's leg, making a little brown rivulet.

My hands drove the horses, but my mind was elsewhere. The noise we made was small—the hollow *clop-clop* of their hooves on the hard dirt road, the rattle of the harness—and the sound of life around us was big, putting us in our proper place. I heard red-winged blackbirds and a nattering squirrel. Then I heard a group of crows putting up such a huge fuss at the top of the hedgerow oak that I scanned the sky for the hawk that I knew must be there, and there it was, a redtail, lifted high over the pasture on a hot breeze. We passed a milkweed plant in full purple bloom, and I stopped the horses for a minute, arrested by its smell, which was as sweet and strong as any pampered garden flower. Why did I feel a quick ripple of sadness? I traced the feeling backward to the fleeting thought that engendered it and found: *I will really miss the beauty of life when I am dead.*

I chose Jay and Jack that day because I didn't need power, I needed slow-paced precision and experience. They were the most patient. Nobody on the farm knew their jobs as well as Jay and Jack, including me. The first task on my list was cultivating the newly emerged carrots, and the first cultivation of the carrots was a big responsibility. Carrots take forever to germinate, and grow very slowly when they do, which allows the weeds to get the jump on them, even in a year when we aren't slowed down by biblical-scale rains. I squinted, looking for the carrot tops among the weeds. The cultivator was fitted with eight shanks and a pair of flimsy metal shields, fixed on a frame that could be steered with my feet and raised or lowered with a lever. The shanks killed the weeds on either side of the row, and the microscopic carrots ran down the middle, directly underneath me. There was a blanket of young weeds around them that were similar in size and only a slightly different shade of green. The straight row of the carrots could be seen when viewed head-on, but disappeared if viewed obliquely, a mirage of regularity in a field of stochastic greens.

We had nine acres of vegetables to weed. In the two days it would take to cultivate, the horses would walk thirty miles. For me, the tiring part was the eyestrain of seeing when the plants were so small. Later in the season, if all went well, the plants would be large enough to brush the seat of my cultivator. The movement of the plants would release their scents, which would reach the open windows of the cars driving by. There would be the tenement smell of crushed onion, the base note of celeriac, and the spicy green aroma of cilantro. But for now, they were all small, and it was up to me to decide how close to come to them. If I were too tentative, I would leave behind lots of weeds that would grow too big to cultivate, pull energy and nutrients from the plants, and create hundreds of hours of handwork. If I were too aggressive, I would bury the tender seedlings and kill them. A few minutes of inattention could bury a whole row. The lesson of cultivation is compromise. You walk into the job knowing you can't kill every weed, and you can't save every plant.

Carrots were always a high priority, because they were a mainstay of our members' kitchens through the long cold winter. If this field yielded decently, we would get eight tons. If the weeds won and the crop failed, we would need to buy them from another farm with money we really didn't have.

Farms teach you that when everything is overwhelming—when you look out at a huge field of weedy vegetables on a hundred-degree day—you must simply pick a row and start. Cultivation was satisfying work. Every step the horses took killed hundreds of tiny weeds. But the most precious part of cultivating that summer was the silence. Outside of the house, alone, I could quietly think. I thought about all sorts of things. Big things and piddly things. The children. Mark and our marriage. The tax bill peeking out from behind the flour bin. My grandmothers. Old boyfriends. Blessings. Problems. Poems I knew by heart. It felt luxurious to think without interruption, to let memories, ideas, or logical connections play out all the way to their gentle ends.

I had not known that we lose the ability to do that when we have little children. Nobody had warned me. And it was a shock. Maybe this was why contemplatives in all those religious orders were celibate: not because of sex itself but for its consequence, the family, which is a beautiful, worthy, but all-encompassing project that tends to march your mind along a trail of quotidian details. Would Saint Augustine have had his revelations if the delicate ember of his idea had been extinguished by a toddler rapid-handling a roll of toilet paper onto the bathroom floor? I thought of nuns and monks in their gardens with their bees. Maybe their lives of chaste fieldwork were arranged to free enough of their minds to give them access to the slippery, elusive divine. I knew that this stage of motherhood would pass, that I would soon miss cool little lips on my cheek, small hands grasping my leg, and the constant stream of needs that I was well qualified to fulfill. I adored my girls so much I felt it physically when I thought of them, a near-pain under my sternum. But for those hours on the gently rocking cultivator, with my two old horses who knew their job so well, it just felt good to be alone.

After noon, I lifted the shanks out of the soil and walked the horses to the trough in the shade of the linden row, at the edge of the field. It would take too long to walk back to the house for lunch, so I'd brought it with me: hard-boiled eggs, a pint of strawberries, a pile of fresh greens, well wilted from the heat, and a quart of water with a splash of cider vinegar, maple syrup, and some salt, our farm version of Gatorade, to keep us vertical when sweat was pouring off us at a ridiculous rate. The week before, Blaine had taken the same lunch while she was out cultivating with the horses, but she had forgotten to salt her eggs, so she had rubbed them on the horses' sweaty necks instead, experimentally, and found it wasn't very good. I took off their bridles, tied their halters to a tree, and sat in the shade to eat. The simple food tasted spectacularly delicious, the way it does when you're hungry.

I lay down under a linden next to the horses, closed my eyes, and fell asleep to the sound of their movements, the creak and clink of their harness when they stretched or shook. I woke up a few minutes later with the image of a painting in my head. Caravaggio's *Conversion on the Road to Damascus*. I saw it in person when I was twenty, in Rome on a summer job, writing copy for a budget travel guide. In the whirl of visiting hostels, cheap restaurants, and museums, I stumbled on this painting. I stood in front of it for an hour, feeding coins to the meter to keep the dim light on, trying to block out all the rococo noise around it. Saint Paul was lying on his back, his hands upstretched. He had fallen off his horse, and God had struck him blind. But the painting wasn't really about the saint. It was about the horse. He was black with large white patches, built tall and sturdy, like a draft. It seemed an interesting choice to me, that Caravaggio had made the horse piebald—why such a flashy ride for a saint?—but maybe he did it to show off how good he was at using white paint. It didn't matter; it worked. In the foreground, a barefoot servant held the horse's bridle, and the horse's front hoof was raised, crudely shod with heavy iron. The saint was almost underneath the horse, and both he and the servant's bare foot looked vulnerable next to the big hoof.

But what Caravaggio put his whole effort into, his whole toolbox for emotion, was the horse's eye. The horse's eye was so gentle. His head was low, and he was watching where he was going. That horse was not going to step on the saint nor on the servant's foot if he could help it. It wasn't lost on me, as I climbed back onto the cultivator, that the saint's conversion happened on horseback, to the rock of a horse's walk. Maybe what Caravaggio was really saying was that vulnerability is necessary if you hope to find some grace.

We finished cultivating the carrots, the potatoes, the sweet corn, and all the greens by early evening. Then we headed back, and I put the tired horses in the barn, unharnessed, brushed, and fed them. I walked toward the house and stopped at the clothesline to unpeg the sheets that had dried to a delicious crispness while I was away. As I folded them into the basket, Jet tore around the edge of the house, and I heard his deep, alert woof. I ran after him and found him taut with importance, his ruff standing up, and the fox cornered under the porch. The fox wore that same poker face, quiet, calm. Maybe he was weak with hunger or the effects of whatever was making him so patchy and scabrous. I told Jet he was a good boy and made him lie down—no need for the good dog to tangle with a sick fox when a gun could do it. I called for Mark, who was upstairs with the girls. The windows were open, and he heard me. He left the kids inside and came out, saw the situation, and limped to the gun cabinet for a sawed-off shotgun that someone had given us once, in payment of a debt—an ugly thing painted flat black that I didn't like to look at, much less shoot. I kept the fox pinned under the porch with my eyes, Jet behind me for backup. The fox blinked but did not move.

Mark came back and took my place. He pulled the trigger, and the sound of the gun filled the evening, its echo rolling over us in waves. Mark forgot to hold the gun tight to his shoulder, and its enormous kick knocked him backward a step. For Mark, that was the only pain of it, but for me, there was that familiar feeling of necessary loss. We can't have a sick fox in the barnyard, killing chickens or hanging around the kids' swing set, and yet who wants to blast holes in a small fierce thing with knowing eyes? I pulled the tattered carcass out from under the porch with my thumb and forefinger. It was nearly as light as the hide and bones it soon would be, and crawling with ticks. Jet jumped around us, triumphant. He is on Mark's team, unequivocal.

I had the chance, during all those quiet hours in the field, to think a lot about our marriage. The truth was, it had been under strain even before Mark got hurt. Our partnership was always as complex as the farm itself. We are both quirky and intense. We are both stubborn, competitive, and generally certain that we're right. Moreover, we have different values, priorities, motivations, and needs. I knew this going in, and so did he. Both of us believed we could make it work, in the same cocky way we believed we could orchestrate agricultural synergy out of our farm's wild diversity.

I suspected that most marriages were more complicated than couples tended to let on. That summer, it seemed like every week brought news of another pair in our community who were splitting up. They were all around our age, most with little kids, trying to run their own businesses in a small economy. The evidence of those breakups directly contradicted the rosy way relationships were portrayed on Facebook, in public. I began to believe that future generations would study how we represent our long-term partnerships, and call us on our lies, in the same way we look at how the Victorians depicted sex and know that it simply wasn't like that, not behind closed doors or in the hayloft. We hide marital conflict with the same sense of decorum. We'd do more good if we were honest and set realistic expectations for what it's like in the long run. Marriage—my marriage—was a long journey across a craggy landscape. Some stretches wound through a green valley. But some were hard climbs up rocks, in the rain. The dangerous parts were the flat empty desert places that didn't seem difficult as much as they were lonely.

The farm was what we had in common. We both wanted it to thrive and succeed. But like any shared project that two people love intensely, the farm also gave us things to argue about. Normal couples argue about kids, money, and household chores. Farm couples argue about roofing the house versus roofing the barn, when to make hay, the condition of the animals, agricultural risks, too much debt versus too much work. Also, kids, money, and household chores.

There was so much we expected of each other, all interconnected. We were house-mates, lovers, business partners, coworkers, and parents together, with each of those relationships demanding a different set of skills, extracting energy from both of us. At times, the farm was so overwhelming that there was no time outside of it for friends, so then we had to be that to each other too. When one of those relationships became disjointed, they all did.

That night, tired from the ongoing battle with Mark, from the farm's constant static, from the mess of our collective lives, I thought, *I might just be using myself up. Maybe I'll die young*. Then, just for a moment: *Oh, that would be so restful.* But the shotgun's kick had jolted something free in Mark's body. He felt a little better, just enough to give him some sleep. And a few days later, as I was coming in from the field, Liz and Brendan appeared, unannounced and most welcome.

They were a pair of benevolent wizards, bearing needles. Brendan was a friend of Mark's from college. Mark described him, back then, as the most radical member of a radical band of activists, on fire for environmental justice, at the front of every march, the first to get arrested. After graduation, Brendan had studied classical Chinese medicine, a form called Five Elements, specializing in the combination of herbs and acupuncture. He had met Liz in acupuncture school, and they'd married, then practiced together in Montana before settling across the lake from us, in Vermont. Liz said Chinese medicine helped moderate the more extreme parts of Brendan's personality, so by the time I met him, he was a proper citizen.

She had brought a bag of surprises for the girls—books, a puzzle, animal toys made of wood. Before acupuncture, she'd been a pediatric occupational therapist, and she thoroughly understood small children. She and Jane were instant friends. We left Mark with Brendan at the house and walked across the farm together, Jane holding Liz's hand, Miranda in a stroller.

Liz was regal, with short silver hair and a perfectly centered manner. She reminded me of a cat. She would laugh—a ringing, free laugh—at exactly what she thought was amusing, but not to be polite, not necessarily at my jokes. Her face seemed extraordinarily still to me, which reminded me how much I had been moving mine around in order to please people, to try to communicate my own agreeableness. Liz's face made me want to stop that, so my face would reflect only what I was authentically feeling. Which was maybe why she could see, in my face, something that made her eyes go soft, her head tilt to the side.

On our way home, we harvested a bowl full of zucchini blossoms, the first of the green onions, sweet tender basil, a handful of garlic scapes, some baby carrots, chard, and a dinner's worth of lettuce. Maybe it was my imagination, but I thought the plants were beginning to look a little bit better.

Brendan was already in the kitchen with Mark, beginning work on a gorgeous frittata, and Liz and I joined them, using our loot to complete the meal. As soon as dinner was over, Brendan and Mark went upstairs for a treatment. We left the dishes on the table while I put Miranda, sound asleep, in her crib. In the next room, Liz read books to Jane, kissed her good night, and turned off her light. "Your turn," she said to me. I sat on the edge of our bed, and Liz sat next to me. She took my pulse, her fingers dancing over my wrist like four little animals with very sharp senses. "Depleted," she muttered, and looked at me seriously. As she gathered her needles, I asked her what she meant. "Look," she said, "a man loses chi through ejaculation. A woman loses it through childbirth. Think of how many times a man ejaculates versus how many times a woman gives birth. You have given so much of yourself in the last five years." I let that sink in as she leveled her cat gaze at me. "You need to do less. You need to rest. You have daughters who need to *see* you rest. Stop jumping up for everything." The first needles were going in, tiny stingers, into my toes, my forehead, my belly. Her voice was hypnotic.

"In our medicine, everyone is ruled by one of the five elements; every element is associated with an emotion. Your element is water. Your ruling emotion is fear and also lack of fear. Mark is fire. His emotion is joy, lack of joy." More needles went in, one by one, and then she left me to let them do their work. I felt drugged, as if the needles were pulling me gently into a peaceful, comfortable room I hadn't visited in ages.

What she'd said about my element and its corresponding emotion was right. I knew it in a place deeper than thought. My fear held me back from full commitment to life as it was, rather than what I thought it should be; my lack of fear propelled me forward, sometimes into opportunity, sometimes into regrettable situations. It was my lack of fear, my ability to leap into the unknown, that had landed me at this crazy farm, but it was my fear that made me feel like running from it, and from him.

Mark's joy was incandescent. It was the volatile fuel that fed the whole five-hundred-acre experiment. His lack of joy, when it hit, was a dark hole that sucked us all in with him.

I thought about fire, and water, and what happened when those two things come together. Liz came back with a stick of *moxa*, made from mugwort, and lit it next to the needles. It's supposed to counter irritation. Before the heat of it had even warmed my skin, I fell into the deepest sleep of my life.

After Liz was finished and the needles were out,

she tucked me into bed. Mark had materialized, tucked in next to me. It was a strange kind of intimacy, not quite private, like Yoko and John in bed together with the photographers around. Brendan and Liz fussed over us, collected their needles, closed their cases. "The girls are asleep," Liz said. "The dishes are done. Stay in bed, we know the way out."

Mark began to get better. He could move again. But for a long time, we didn't move much closer together. When he was well, he was fully well. He got out of our bed, returned to his old grueling hours with the same old intensity. It was like the rain: because the pain and darkness had ended so abruptly, they seemed, in retrospect, like a bad dream. He didn't seem to remember how bad things had been. His eyes were fixed, as usual, on the future. He began running the farm again, zero to a hundred in the course of one week. His old spirit returned, along with his desire to manage things down to the smallest detail. His drive to build the business into something secure doubled, now that its existence was threatened. He reminded me of how he looked after windsurfing. Happiest when he was clinging to the mast, just this side of disaster.

He tackled the finances first, pulling the tax bill from behind the flour bin. There was a flurry of phone calls, a long discussion with an accountant, some papers to sign, and then the bill disappeared. Someone had ticked the wrong box, the whole thing was a mistake, and we didn't owe the money.

Then he turned to the fields. We'd never used anything but our own compost to feed our soil, and we'd relied on it to keep the plants healthy enough to grow and fend off the pests, but this extraordinary year demanded something different. Mark dipped into our credit line to order five tons of organic fertilizer to replace the nitrogen that the rain had washed away, and used it to side-dress the peaked plants. They sucked up the nitrogen, spread their leaves in the summer sun, and grew.

And I went back to the house with the children, because that was where I was most needed. Our farm kept trying to separate Mark and me into traditional gender roles, and I wasn't sure how to feel about it. I had been raised in a family with a full-time homemaking mother right at the end of the era when that was default normal. But I was the cultural product of third-wave feminism and had assimilated all the ideas. I had married someone who was enlightened enough to take my last name. I never doubted that I was a full and equal partner in our business. And yet I was the one who was mostly in the house, at the stove, taking care of the children, while he was running the business, managing the employees.

It wasn't even much of a discussion between us. *It'll probably even out later on,* I thought. I kept my milking shift, and for a while, that was enough to make me feel like I still had a role on the farm. However, when Miranda had grown heavy but hadn't yet learned to walk, I blew out my left shoulder from carrying her on my hip all the time, trying to do everything right-handed. It took months to heal and made milking impossible. I couldn't lift the ninety-pound milk cans or tip the buckets into the filter. The irony of it—that motherhood was more of a strain on my body than farming—depressed me.

I'd had a lot of ideas about myself and work and motherhood before kids, and they didn't match up with the reality, and that was difficult to reconcile. I half-consciously believed I should have been capable of giving efficient birth in the hedgerow, picking up the baby and a hoe right afterward, and continuing along. Once a young woman came to farm with us for a week, bringing her fourteen-month-old daughter, who rode around with her in a backpack. The woman and her baby spent a long day in the field, weeding, fencing, harvesting. Just before dark, they came back to the barn, the baby slightly toasted from the sun and bitten by gnats. The mother pulled out some farrier tools and trimmed the hooves of a team of horses—the heaviest kind of awkward, muscular, bent-over work—while the baby slept on her back. I thought I *should* be able to do that too.

It turned out I couldn't. Not that I didn't want to, I just couldn't. Maybe, I thought, it was because I was a geriatric multipara. That woman was almost twenty years younger than I had been when Miranda came along. The thought hurt my ego. In my family we did not blame age for weaknesses. My father took up kitesurfing in his seventies, skied black diamonds well into his eighties, and I was cracking under the weight of a mere baby?

For the rest of that year, Mark and I divided our labor for the good of the whole farm. The children plus the cooking made a full-time job. The farm was full-time and then some. We could have split those jobs between us, but I didn't have the same set of skills that Mark had. I wasn't as strong. I was roughly half his size. It was irritating, but he had a decade more farm experience than I, had apprenticed with master farmers, and had a degree in agriculture. I was an English major who had spent my twenties with books in New York City. I didn't know how to weld or fix an engine or do basic carpentry. If I broke an implement I was using, I'd need to call someone else to fix it. I didn't know the names of half the tools in the machine shop.

Mark was as much of a menace in the house as I was in the shop. He wasn't patient with the details of the housework and children. His love of efficiency made him want to pare down those duties to their most minimal forms. Once I found him sketching a baby-shaped bag in his notebook. He'd had enough of the fussy little socks, pants, and shirts that had to be put on and taken off a wriggling child, washed, dried, and paired up again. So much time! So tedious! Too many holes! It would be more efficient, he said, if we dressed both of them in one-piece *sacks,* color-coded by child, and had seven *sacks,* one for each day, no extra pieces. "Why can't we?" he asked, dead serious. "Because they are humans," I answered. "They are small but real people, and people wear clothes."

During his recovery, when he was well enough to get out of bed but before he was strong enough for farmwork, he'd developed a theory he called "micro-parenting," which stretched the limits of how little one could do on behalf of a child and still keep her safe and generally happy. This style had nothing to do with how much he loved the girls. He loved them as I did, entirely. But it was related to his desire to spend every moment as though it were his last precious coin. "What would happen," he wondered out loud at one point, "if we put the children in an electronet like we do the chickens? They could be outside, run around all they wanted to within it, but stay safe from cars, tractors, and horses. I bet they'd learn the boundary faster than a chicken does. Maybe get shocked once?"

"What would happen is," I said, "the authorities would come and take our children away and put you in a cell."

Often micro-parenting created a huge mess. He didn't see the value of folding clothes or putting them away. Why not just heap the laundry, but in a kind of spread-out heap, so you could see what was there and grab it? I'd come in to find Jet sleeping happily on top of the clean clothes, shedding hair, mud, and straw into them. Mark had calculated, in his head, the amount of time we spent washing dishes; he floated the idea that we should eat directly off the table instead of plates, then clean up with one swoop of a drywall scraper into the compost bin when we were finished. Why fuss over socks or dishes when we could be doing something better together? Even if that was reading a book or playing tag? Weren't those things more important? And underneath that was the feeling that if he relented and began to care about the details that other people cared about, he'd be overwhelmed by the enormity of our farm, and it would die.

People who came to the farm during that period saw me as a different person from who I thought I was. It hurt in the weirdest way. To these young newcomers, I was a mom, middle-aged, with tired-looking breasts and sun-damaged skin, and therefore, mostly invisible. I wasn't the farmer but the farmer's wife.

One Friday, our friend Matt came to visit and stayed for Team Dinner. Seeing him was always a treat. He ran his own farm in a different part of the state, a diversified CSA, and draft horse–powered, like ours. He'd come to work for us part-time in our second year farming, when he was a college student across the lake in Vermont. If our farm could have invented for itself the perfect person to help us work it, it would have come up with Matt. He was a star rugby player who didn't drink, a philosophy major who was grounded by physical work, a heavy-metal fan who went to Mass every Sunday. In other words, tough, strong, serious, intelligent, fun, and thoughtful.

One of the biggest compliments in Matt's vocabulary was "savage." His best rugby teammates were savages. The drummer from Pantera was savage. Matt quickly fell in love with farming, and nothing we threw at him could shake him from it. He drove the hour from his campus to our farm after classes some days, and he spent most weekends and vacations with us. When he graduated, he came to work for us full-time.

Around that same time, we hired Sam. Mark had known Sam since he was a little kid, the much younger brother of a good friend. Sam had just graduated from college and, like Matt, was smart, strong, athletic, and soon in love with farming. We four made a good team. Mark and I taught them to harness and drive horses; butcher cattle, pigs, and chickens; and milk cows by hand. They were eager to learn, and they worked themselves to blissful exhaustion. They built their own cabins to live in, next door to each other in the woods along the farm road, surrounded by cedar, sumac, and wild blackberries. They had no electricity or running water, but each one had a small woodstove inside, and lots of insulation, and were snug and comfortable even deep in winter.

It was a beautiful thing to watch those two work. When I think of that time on our farm, I see Matt and Sam running through the hayfield at dusk, the remnants of August heat coming out of the ground, throwing fifty-pound bales onto the horse-drawn wagon that I'm driving. Mark is behind me on the swaying wagon, stacking, chaff stuck to him, his long work-hard arms swinging the bales above his head as if they're bits of mere fluff. There's the good green smell of the new bales and the sweat of the horses and the dusty scent of the hot raked earth. The grasshoppers and crickets are jumping in front of us like popcorn in the stubble of the cut grass, and Jet, in his youth, pouncing on them. The red-tailed hawk is circling above, scanning the newly bare ground for voles and mice. Matt and Sam are singing loudly as they run, "Fox Went Out on a Chilly Night," making up new verses. They run all evening from bale to heavy bale, jumping over the tongue of the moving wagon because they are young and strong, and they can. Those were moments of pure beautiful happiness, a crystalized image that combined all the best parts of farming, which is, in the end, a vigorous team sport.

We were hosting a big dinner the night when Matt came to visit. There were twenty people eating, a group made up of volunteers and farm visitors in addition to our full-time crew. Everyone except me had spent the day in the field, haying, harvesting, weeding, moving animals to new pasture. I'd been inside most of the day with the kids, cooking the enormous dinner, hauling several loads of laundry from the washer to the line.

All the leaves were in the table, and still it was too crowded, people spilling onto the grimy couch or the piano bench, or perched on folding chairs against the wall, plates balanced on their knees. Passing between the kitchen and the table with a platter of food, I heard a newcomer ask Matt what it had been like when he'd worked here.

“Well, it was just the four of us,” Matt said, raising his steady voice to make himself heard above the general roar. “Sam, Mark, Kristin, and I. We did everything.”

“*Kristin* used to farm then?” the newcomer asked, glancing up, incredulous.

Had I *farmed?* My mind shot to one particular day during haymaking, five years earlier, when Matt and I were in the loft of the East Barn, stacking bales in hundred-degree heat, the bales coming along the overhead conveyer at a fast clip, crashing to the barn floor, the hay dust floating thick in the soft light and sticking to our sweating skin. If we slowed down and got behind, the bales would pile on the floor of the barn or, worse, come down on our heads. The gritty good feeling of hard exertion and teamwork shifted to a sort of gasping, determined desperation. I’d tried to match Matt’s pace and keep up with the relentless bales, and for a while I did, and then I suddenly had to stop, double over, and throw up. We yelled to Mark to slow the pace of the bales a little, and then we finished the load. *Yes, young man,* I thought, *I farmed then*. I wondered if Matt remembered that day.

“Kristin?” Matt answered the newcomer. “Oh yeah, Kristin farmed. She was a savage.” He remembered. I smiled to myself and cleared away the dirty plates.

CHAPTER 10

Mark still valued my input, we discussed decisions both large and small and strategized together every day, but our realms were clear. He was running the farm, and I was taking care of the children. I loved caring for the girls, especially when the demands of the day were only to keep both of them safe and fed, and give them plenty of the loving eyeball-to-eyeball attention that helps knit their neurons together. But it was isolating and frustrating to watch what felt like the real work take place outside, without me.

My eighth year in Essex I packed up my ambition, all the energy that I was used to spending in the fields, and brought it to the farmhouse kitchen. Before farming, I'd liked food in the same way I liked all sensory pleasures, especially novel ones, but I hadn't been much of a cook. I'd lived in a six-hundred-square-foot studio apartment with a tiny stove and inadequate ventilation. I was single. Nobody in my group of friends cooked much. We all believed vaguely that cooking would be for later, when we'd have bigger apartments, maybe partners or families, and would have become "people." We didn't know a lot of people, but we could see them making dinner in the evenings through the windows of their brownstones. We thought of ourselves, by contrast, as something a little less substantial. And until we became people, there would be restaurants, and takeout, and no dirty dishes.

Mark had seduced me with his cooking. He'd been cooking all his life. He started farming in order to access the quality of food he wanted to eat. The meals he made for me were a lot like his personality: big, sure, and deeply, deliciously strange. Deviled kidneys, curried parsnips, a rib roast of beef dry-aged in the cooler until the outside was covered with a soft white fuzz that made the meat taste like mushrooms. Sturdy, nourishing soups, sour with fermented vegetables. Salads of shredded celeriac, a mess of bitter wild greens. Squash pie with star anise and maple syrup, encased in a flaky lard crust.

When we were first married, we cooked together all the time. What else were we going to do through the long dark winters when we weren't outside? We didn't have a TV, we were often too tired to read, and we were surrounded by vast amounts of fresh, delicious food. Moreover, we were always starving because of all the hard work. In those early years of cooking together, I absorbed some of his skill at preparing whole food in season and added to it my own interest in flavors from other cultures, plus deep dives into books by food writers and chefs, from the tart and scientific *Cook's Illustrated* to the writerly narratives of Jane Grigson and Laurie Colwin. And then I practiced until I was as confident and efficient if not quite as wildly imaginative in the kitchen as Mark was.

After the girls were born, Mark stopped cooking almost entirely, and I took over. As with our other roles—housework, child care—we didn't explicitly talk about the shift. It just slowly happened. Sometimes I missed the strange, seductive dinners for two, and our shared hours at the stove, but he was always out in the barn or the field, so busy, and I was inside most of the time anyway, with the kids. And cooking for all of us every day made me good at it. I learned the importance of forethought. If you are going to soak beans or make bread, you have to think at least a day ahead. For everyday meals, I learned to moderate my ambition and aim for the large territory of great food at the intersection of easy and delicious. I also realized that cooking never failed to cheer me up. The joy I took in our food and in cooking was immune to the ups and downs of the weather, the season, the finances, and our marriage. And it could be done safely, sometimes even gracefully, with children.

When the tiny row of cilantro was tall, I took the girls with me to harvest all nine hundred feet of it, enough to supply a sizable city. We should have planted it in successions, a hundred feet every week, but sometimes, with our horse-drawn systems—and especially in a year so foul—it was easier to plant a whole row than to plant a section, so we ended up with extreme excess. It would bolt soon, and that would be the end of cilantro for the year. Looking at it from the edge of the field with the girls, I got an itchy, covetous feeling. An abundance of herbs, like an abundance of flowers, still made me feel rich. Jane and Miranda played between the rows as I harvested armloads.

When we got back to the house, I washed out the field heat and the sand in the cold water of the deep kitchen sink, then hung it in a mesh bag outside on the clothesline, so the afternoon sun and breeze could carry off the excess water. Then we laid the green loot directly on the kitchen table, which doubled as an enormous cutting board, and I went at it with the big kitchen knife, rocking back and forth all the way down the table until the cilantro was roughly chopped. Then I pushed handfuls of it into a half-gallon jar, added olive oil and salt, and pulverized it with my immersion blender. I froze the green paste in ice cube trays. The next day, I would pop out the cubes into gallon freezer bags and haul them to the chest freezer in the basement. No matter how much cilantro I froze, I always wanted more, because the smell of it has the power to transport me around the world.

Travel was the love of my youth. After the farm, marriage, and children grounded me, I transferred that love to food and cooking, but they were related. I had grown up in a small town in central New York and gone to a big public high school, and that first year away at college gave me a glimpse of the breadth and depth of parts of the world that I hadn't known existed. All of it was intriguing. I wanted to read everything and do everything and taste everything in gulps, and test the boundaries of who I could be. I was unsophisticated and unsure of myself but curious.

I started dating someone in my class who had a Turkish father and an American mother and had grown up in Istanbul before going to boarding school in Massachusetts. He kept a finch named Sufi in his dorm room, and when he spoke Turkish to his father on the phone, I drew close, trying to tease meaning from the rolling syllables.

He invited me to visit him in Turkey that summer, to meet his family and see where he was from. After my last exam, I got a cruddy apartment in Somerville, took wretchedly dull temp jobs in Boston, and saved every penny until I had plane fare and spending money. Just before I flew out, my sister's then-husband put a hundred dollars in my hand and told me to go to the bazaar in Istanbul and buy something for my sister. He understood me and knew that the gift—which was really the imperative to explore—was the best thing he could have given me.

I don't remember the airplane leaving Boston or landing in Istanbul. But recollection begins crisp as a movie at a table, the first meal. We'd braved what felt like life-threatening traffic to get there. In the calm courtyard of the restaurant, there was the smell of the Bosporus, the warm sun, a perfect white tablecloth. My boyfriend and a bent old waiter exchanged words that conjured a dish in front of me. It was pasta, but unlike anything I'd ever tasted—a delicate ravioli with a warm garlic-yogurt sauce and a drizzle of bright red butter infused with paprika. In my hometown, pasta invariably came with tomato sauce, and yogurt was mixed with fruit, always sweet and cold. Savory, hot, garlicky yogurt was a revelation, and I asked for the name of that dish—*manti*—so that if I got separated from my boyfriend, I could order it again. There were other new flavors and good meals on the trip, but that first one awakened me to a vast new possibility of taste combinations.

The summer after sophomore year, I got a job writing for a student-run travel guide, and after that, I was off on a long string of trips that continued until I met Mark. I had favorite places in Hawaii, Thailand, Indonesia, Rome, Mexico, Brazil, Spain, and Burma. Sometimes I traveled alone, working for guides or magazines, and sometimes with my sister, on exploratory trips that we planned and saved for over the course of many months.

As the years on the farm rolled by, I could feel my grip on the details of those places slipping, the maps and names of all the cities I'd visited growing less legible, but my memory held tight to the smells and flavors. Hawaii was ahi poke and rice with kimchi on a picnic table at Ho'okipa while the breakers crashed in salt air. Indonesia, with my sister, was grilled tempeh, gado-gado, fiery sambal, and the smell of clove cigarettes. When I chopped cilantro in my kitchen, I thought of the Mexican highlands where I'd lived for over a year, and the bundles of herbs piled in the market in San Cristobal. The rest of the market was there with me too: the mixed smell of chilis, rosemary, animal blood, woodsmoke, and fresh handmade tortillas with the wool of the indigenous women's skirts, which were damp with mist.

Farming grounded me. I wouldn't be traveling very far when tethered to a piece of land and two small children. Under those circumstances, it wasn't even easy to go to the store. But I was surrounded by the best-quality food, and I could cook and remember. Those years when the girls were very small, the three of us spent a lot of our day together in the kitchen, and went deep into cooking projects. We'd read about them, look at maps and pictures, and then begin. The kids and I harvested and nixtamalized corn, ground it with a molino, and pressed the masa into thin circles to make tortillas; we fermented daikon, cabbage, and garlic into kimchi; we made tempeh and miso from our soybeans. We spent weeks on Persian food, with prodigious amounts of fresh herbs, and on the different cuisines of India, and Jane's favorite meal became saag paneer made with our milk and yogurt. In order to expand our reach, I bought what we couldn't grow: fresh ginger, olive oil, wine, rice, and citrus in bulk quantities. I kept my spice cabinet fresh and full, and lined my pantry shelf with cookbooks from all the places I wanted to visit or revisit. That collection was our ticket around the globe and made cooking and eating whole seasonal food interesting and delicious all year.

Even when a farm is struggling, even when there is no money, there is always plenty of good food. After we kicked the farmers out of the house, I focused more energy on our weekly Team Dinner. This was one of the traditions that we'd started early on, when we had our first employees. Every Friday night, after our members had all come and gone with their cars and trucks loaded with food, everyone who had worked on the farm that week was invited to gather at our house to celebrate. Team Dinner included full-timers, part-timers, and also alumni, significant others, children, visitors, extended family, and any stray wanderers who looked like they could entertain us. Agricultural theories, riotous jokes, lifelong friendships, and romances took root around the Team Dinner table. Once when my sister was visiting, she looked around the table at the strong and nourished people and suddenly felt like she was missing out on something in her city life. She suspected the rest of us were in on a form of delight that she couldn't quite access, even in a city that could sell any pleasure a person could imagine, at any hour. Most weeks, we thought so too, secure in the bounty around us, and the specific sort of hard-earned joy that farming gave us.

Part of the enjoyment was due to the fact that everyone worked hard and was hungry. Part of it was because all the ingredients came from the good soil around us, often harvested minutes before hitting the pan. And part was the simple beauty of eating the result of all the work that we did together. The people around our table came from different backgrounds, were different ages, and had different beliefs and experiences. At one point, our farm crew would be a combination of evangelicals, Old Order Amish, secular Jewish people, atheists, plus one ex-Amish messianic Jew—a mix Mark called the most diverse set of white people ever assembled—but the table always provided common ground that made us agreeable, sociable, and willing to transcend our differences.

Some nights, Team Dinner was simple—a giant pot of stew and a loaf of homemade bread—but often I aimed for something more festive and, occasionally, a blowout feast with several courses and a unifying theme. Always, it was built around whatever food was perfect at that moment in the season.

The rest of the farmhouse—the peeling wallpaper, the dim lighting, the fake-wood paneling in our bedrooms—depressed me. I seemed unable to do much about it. There was never extra time or money. Mark disliked anything that distracted us from the farm and its needs. He made it clear that he would not be doing any home improvements. Nor did he want to spend money for someone else to do them. If I attempted to fix something on my own, he pointed out the inadequacy of my skills. When the washing machine broke and I wanted to shop for a new one, we had an extended argument about the efficacy of doing laundry in the bathtub with a pair of plungers.

But the kitchen was an island of perfection in that sea of ugliness. It had evolved, by necessity, into something hard and purely functional, which gave it a utilitarian beauty. There was a nail-studded two-by-four at eye level on the kitchen wall, for hanging our cast-iron pans. The large kitchen table that doubled as a cutting board was set on bricks to bring the surface to waist level for ergonomic chopping. Our sink had been salvaged from a restaurant and had three enormous bays, one for spraying, one for washing, one for rinsing, and took up nearly one entire wall. Over it, two parallel metal pipes were driven into the walls, supporting a twenty-foot-long rack made of chicken wire. This was where we stored the large pots and pans after they came out of the sink. Three people could wash dishes for a table crammed with farmers and have them wiped and put away in ten minutes.

Sometimes I wonder if the fruitfulness of

that period of our farm—the way it spawned couples, other farms, and babies—was related to the elaborateness of those Team Dinners. I watched the evolution of romances that summer, all around the table. Racey was in love with Nathan. Tobias was in love with Blaine. Chad had just met a newcomer, Gwen, the woman he'd eventually marry and build a farm with.

Sometimes I planned Team Dinner well in advance and started cooking days ahead. Other times, the meal came together all at once, through inspiration, when I caught sight of creamy white cauliflower coming ripe in the field during a morning walk, or an unusual cut of meat hanging fresh in the butcher shop. Jane spent some mornings that summer in front of the butcher shop, sitting on top of an overturned bucket, watching Blaine eviscerate pigs and steers. Jane's fascination with this process was due in part to her attachment to Blaine, who could be acerbic with adults but was invariably attentive and kind to Jane. Blaine had given Jane some of her favorite books from when she was a child: *Harold and the Purple Crayon* and *Happy Birthday, Moon*. Those books became Jane's favorites because they'd come from Blaine.

As Blaine butchered, she chatted with Jane and narrated the work she was doing. She showed Jane how she slit the animal's belly skin, the knife held at its hilt with the sharp edge pointing out, being careful not to puncture the intestines. She demonstrated how to saw through the breastbone, tie the bung, and enjoy the suspenseful moment of repose just before the innards fell out in a heap. She pointed out the vast white bag of the rumen, which held so much fermenting green grass, and the heavy, limp bloodred liver, which she rinsed and impaled on a rack of spikes in the cooler to chill, and which we might have later for dinner. Jane watched it all like television, entirely absorbed, for long stretches of time. By the time Miranda was old enough to venture away from me, Blaine was gone from our butcher shop and on to her own farm, but Miranda and Jane would still dash outside when they saw the tractor roll by our house with a carcass swinging from the raised bucket, to see the action.

One morning I went out to find Jane talking with Blaine over a pair of pigs that she was butchering. Blaine was laying open the body cavity of the second one as Jane peered inside. The arrangement of the organs is a divine packing job, no space wasted and nothing extra. It's finished with a drape of strong membrane that is so thin it's translucent, a flexible windowpane run through with lacy veins of white fat. This is the caul. Jane was poking at it, curious. I took it from the steaming cavity, along with a kidney, a hunk of liver, and the heart, and trotted back to the house, suddenly sure what we would have for Team Dinner: *crépinettes* in French, *fegatelli* in Italian, and genius in any language, which I'd read about but never tried to make.

I pulled my grandmother's meat grinder out of the drawer, a pocked old thing that made me happy to use, like her scissors from the shirt factory, because I could feel the ghost of her slim strong hand on the handle. I'd carved a notch out of the bottom of our kitchen table to attach it, because the wood was too thick. I chunked the organs into pieces and mixed in hunks of stew beef and pieces of pork shoulder. All organ would be too intense. All pork, too fatty. All beef, too lean and boring. A third each? Perfection. I ran to the garden, knife drawn, and hacked off handfuls of sage leaves. Back to the kitchen to put it all through the grinder. Then Jane cracked half a dozen eggs for me, and I mixed them with the ground meats, sage, fresh cream, some bread crumbs for binding, plus salt and ground black pepper. When it was all mixed, I fried a sample for us to taste, adjusted the salt and pepper. Then the fun part. We cut the caul into squares and wrapped balls of the mixture in it, like little meat presents, and tucked them into a hotel pan and popped them in the oven. The heat shrank the caul and melted the fat around the meat into something so juicy and delicious-looking they came out at Team Dinner to actual applause.

• • •

Despite the difficulties, our membership was growing quickly, by 10 percent per year. Employees were necessary, but they brought a different kind of work, less sweaty but more stressful sometimes than plowing or weeding. The accounting became more complicated. Mark had always done the books old-school-style, by hand, in a ledger, with a pencil. Now we needed more efficient systems and different tools. We switched, laboriously, to using a computer. The same thing was happening all over the farm. One piece of infrastructure would get too small to work efficiently—the grain mill, the walk-in refrigerator, the root-washing system—and need replacing, and then another and another in a constant stream of expensive improvements. Mark could manage it most of the time. He's comfortable in a near-crisis, putting all his substantial energy into an immediate and urgent problem. But for those of us not built like he is, the plotline could be exhausting, as if watching a movie made up entirely of climaxes. In one argument, he said to me, "I can't live with you if you can't bounce back from the chaos I create." It took years, after I heard that, for me to realize he was telling the truth.

The economics of using draft horses were becoming harder to manage as the farm grew. As more of the work was done by employees, on payroll, the cost of using horses began to feel untenable. Time was money. It cost money to harness the horses, to care for them, to feed them. It cost money and other opportunities to work at their slow pace. And it cost money to train people to use them. We could spend a year teaching a new person to drive horses, slowly increasing the complexity of the jobs he could do, the number and type of horses he could handle. Then he might stay another six months or a year and move on, to another farm or his own place, and we would have to start all over. Yet Mark had a vision of thirty teams in the field, fifteen teamsters, everyone parading out of the barn at dawn for a day of beautiful sun-powered work. Buying diesel for the tractors pained him, not because of the expense but because he could see firsthand what climate change was doing to the planet, and he knew we were part of the problem. He stubbornly wanted a better way.

As much as I loved the horses and agreed with his environmental concerns, I couldn't see how it could work, as we grew, to keep them as our main source of power. We'd do no good to anyone if the farm was insolvent. "You're *delusional,*" I said to him one night. Our arguments almost always happened at night, when we were exhausted. In the old days, we could work those tensions out in bed, but these days, with one child in my arms and another nursing, the last thing I wanted at night was to be touched. I'd spent all day being touched, mouthed, and made unpleasantly sticky. I wanted my body to myself. Which didn't help the mood between us.

"I'm not delusional," he answered, looking squarely at me.

"Ninety-nine out of a hundred people would say that you are," I said.

"Just give me five hundred years," he said, "and they'll all say I was right."

"You'll be dead! We can't do what's impossible here and now. We have to choose. We can be small and weird," I said, "or big and normal, but we can't be big and weird."

"But I am big and weird," he answered. And I couldn't argue with that.

Scale is just scale, but it's everything in farming. Scale dictates the equipment you use, the infrastructure you need, the labor you employ. When the size of the farm pushes up against the limits of your current scale, things go awry. The well, for example. The final crushing irony of that wet, wet year was that our well went dry. It was an artesian well, and the farm had grown to the point where the natural flow it provided was not enough to reliably keep the stock watered, the dairy equipment washed, and the chicken carcasses chilled on slaughter day. For a while, it was a problem only when we had a leak in one of the miles of hose that we fed out along the pasture. The leak would drain the cistern, and when I woke up to make my coffee, I'd find the kitchen faucet dry. There would be a panic to find the leak, to get water to the thirsty stock. After we found the leak, fixed it, and let the cistern refill, the water would be full of fine sand and silt, which would wear down valves all over the farm, leave grit in the bottom of the teakettle, and break the insides of my hard-won household appliances.

When the weather turned hot and dry, we started running out of water even without leaks. I knew that we needed to do something about the well, but couldn't see where the money was going to come from. It was an inflection point, another chance to disagree about finances, about our identity and what we were supposed to be.

There was a limit to how big and homogenous a farm could be and still be a place where I wanted to live. Large-scale monoculture is very efficient economically, but even within organics, it tends toward systems that work against, instead of with, the rules of nature. And farms like that do not feed people directly, not even the farmers who work them. The product has to be expensively manipulated and packaged into something else before returning to the table. I had no interest in growing five hundred acres of carrots, or milking fifteen hundred cows, then buying my vegetables or milk at the store. Mark and I both knew that a farm like that might solve our financial woes, but it wouldn't make us happy. We could also see there was an opposite limit to how small and diversified a farm could be and still provide a living for a family, at least in a world that requires property taxes, a mortgage, health insurance, and shoes.

Some farms find their sweet spot by getting very small and specialized. We gave a talk at one farmer conference with a couple who made a very respectable living on a single acre of high-end vegetables, sold mostly to chefs and at farmers' markets. They grew fancy produce in greenhouses, under plastic, where they could control the weather with heaters, fans, and sprinklers. They had no children, no employees, no farmhouse, no mortgage, no land of their own, just two miniature dachshunds. They had barely, it seemed, any dirt. They showed crisp graphs and spreadsheets during their talk, which supported the intelligence of their decisions. It sounded perfect for them and perfectly awful to me. Even though it made me crazy, what I loved about our farm was its wild diversity, its mix of plants and animals. I loved the enormous compost pile that took in all forms of life and cooked it into live food for the soil. I loved the soil itself, and the richness of its biology, which was the sum of all our input and decisions. I believed that was the origin of our magic. And I knew that I loved the food. Baby vegetables were beautiful, and apparently quite profitable, but they weren't what filled people's bellies every day. That job was for fats, proteins, and carbohydrates, which came from elsewhere.

There must have been a significant part of me that was drawn to the gamble too, and to our long narrative full of climax. I didn't want to live and work under plastic, even if it kept the rain off my plants. The risk and the unknown were part of what made it attractive. I had, after all, sidled up as close as I dared to Mark's incandescent chaos. I'd married it.

Mark and I weren't disagreeing, exactly. We were trying to sort out what we were, and Mark's way of knowing what he thinks is by arguing it. He'd ask what seemed like an open question, then pounce on my answer and worry it to the ground. That made me want to defend my original position with my life. Our evenings together, which used to be steeped in sex and food, were now full of loud debate. What we were talking about was our identity. I was feeling the precariousness of holding on to what we had made, which I loved as much as I loved him. It was dizzying. In the big stack of bills and the new line of credit, I saw a hard truth: the farm that had been so hard to build would be so easy to lose.

We would need to buy almost all of our grain that year, because the rain had not allowed us to plant any. Our hay was made late in the season, so it was plentiful but poor in quality. We'd probably need to buy some good hay to get us through the winter. In bed, in the middle of the night, I mentally weighed our income against our expenses.

We hadn't expected the farm to give us a middle-class living. After a few years, even I, raw beginner, could see it wouldn't. Mark and I promised to take each other for richer or poorer. That was a trade we made for farming the way we wanted to, with only a third of our attention to the bottom line, the rest of it focused on the quality and diversity of the food, the health of the soil, the effect on the environment and the community.

Once, as I was walking down the farm road to bring the cows up for milking, I was arrested by the sound of a violin. Not a beginner's squawk but a fine full tone, two notes in harmonious conversation. I stopped and looked for the musician. It was the limb of an arborvitae, grown close around the wire of the fence and played by the wind. Farmers like us are always poised between nostalgia for a past that never existed and hope for an idyllic future that never comes. The tension between those two narratives is part of what keeps us in it, vibrating between them like a note pulled from a taut string. In the beginning, when Mark and I started the farm, the past was the image that got me out of bed to do chores every morning. I thought we could create happiness by reinventing the farm of our grandparents' generation—by working hard together on something that made sense to us and fed us on all levels. After the children came, it was the future that motivated me, the hope that we could make a good and modestly prosperous life for them out of the sun and the dirt. I wanted to build something beautiful and secure enough that they would want to inherit it and continue.

Ben Christian is a generational farmer. His grandfather farmed, his father too, but by the time he and his brother were grown, the land had been sold; the old farmhouse, on a road named for their family, had been sold to out-of-towners. But Ben and his brother, Scott, had spent their childhoods in barns, around cows, with 4-H, or working with extended family. Farming was in them, bone-deep.

Ben is large, strong, the agrarian equivalent of street-smart—field-smart, maybe—and also an excellent storyteller. When I had a problem with a dairy cow that I couldn't deal with on my own, like a ripped teat that dripped milk, or a knocked-off dehorning scab spurting arterial blood into the air, Ben was my first panicky call. If he could come, he would, often with Scott and Scott's son, Jon, along for the ride. Scott and Jon knew a lot about cows too, but Ben had managed a thousand-cow dairy for a while, and that, combined with his lifetime of experience, meant there was not much about dairy cows that he hadn't seen. He approached any problem in a cow or elsewhere on our farm with the same level of acceptance, no matter its size or complexity. This isn't to say he didn't yell or curse. But he had achieved a hard-won detachment, a tacit agreement with himself to take what the world or a cow or the weather or the bank could throw at him, be it beautiful or evil. When it was the latter, he'd quote his mom, a farmer's daughter herself and later a farm wife, who told him that if farming were easy, everyone would do it. He might groan like the rest of us when something went wrong, but there was no panic in him. Except when it came to snakes. He was afraid of them, and because of that, he liked to tell stories about them.

Venomous snakes are practically nonexistent in the Northeast, with the exception of the timber rattler, a threatened species that lives in a few very small, localized pockets in our region, where conditions are just right for them. There is a place like that south of us, close to the lake, in a section of wilderness called Split Rock. Timber rattlers are rare in the big sense, but in their little just-right spot, they are quite common. People find them in their garages or yards in the summer; there are signs around Split Rock warning hikers not to stick their hands into any dark crevices.

Split Rock is surrounded by farmland that has been worked for generations. Ben knew each field and all its details: how much the owners had paid for it, how well or poorly they had managed it, and when it had changed hands. In Ben's grandfather's time, the farm next to Split Rock was a dairy. I picture a garden for the family, canning for the pantry in the fall, some chickens, a pig, sufficient good food for everyone.

Now the timber rattlers are carefully protected, but back then they were considered a public menace, and there was a bounty on them, two dollars per rattle. The state didn't care how big the rattle was. So when the summer was coming to an end, that old farmer would wait for a cold night, pull on his tall boots, get his shotgun out, and hunt for the rattlesnakes while they were torpid, in their dens. He was looking for nests full of hatchlings. One blast of shot might net him twelve or fifteen tiny rattles and maybe one big one, and he knew where lots of the dens would be. This was the money the family used to buy the children's shoes for school in the fall.

What I heard from the story that summer was different from what I would have heard before having kids. It didn't sound like a nostalgic story of rural resilience to me, of old-fashioned

American self-reliance. And my first thought wasn't outrage on behalf of the poor baby snakes. Instead, I heard the stress and worry the family must have felt at not having enough cash to put shoes on their kids' feet. The shame at sending your children off to school without the things they need. I heard the absurdity of having to use dead rattlesnakes as your currency when you are employed, full-time and then some, to produce food for your fellow humans. I heard that the cash gain from throwing oneself so hard into difficult work was sometimes not enough to keep the family afloat.

CHAPTER 11

Mark was alive with his usual physical ambition, his optimism and focus. He took the dire circumstances and shifted them into action. "If this is community-supported agriculture," he said, "then let's reach out for some community support." He had never been shy about asking for help for the farm. With a long stretch of dry weather at the end of July came an enormous cascade of work. One daunting weekend, we had fall transplanting and weeding that had to be done, plus a thousand bales' worth of hay that had been cut and would need to be tedded, raked, baled, and stacked in the barn before the next rain.

He put out a call for volunteers of any kind. I listened to him working the phone, going through his whole list of contacts, looking for experienced tractor drivers, able-bodied hay movers, auxiliary kitchen support, and child care for Jane and Miranda. When he didn't hear back from enough people, he sent word out again, louder and farther, until we had gathered a dozen neighbors, friends, and relatives, along with our own crew. Everyone set to it in the July heat, and by the end of the weekend, the hay was in the barn, the transplants were set in the field, and the weeds had taken a serious hit.

I had Miranda in my arms and Jane trotting along at my heels that summer, and what I wanted—what I *needed,* I implored Mark—was a *reasonable* sense of security, some sort of guarantee that we'd never see a season as difficult as that spring had been, ever again. I couldn't shake the memory of watching the rain pound against the window and fill the puddles in the driveway as we sat helplessly inside. And if we were being totally honest, it wasn't just that spring. The rains had been extreme beyond anyone's living memory, but it was only a more intense version of the problem that had plagued us from the beginning. It had been the third spring in which too much rain had made our planting late or difficult. Of the eight springs we'd seen on the farm, our biggest problem for seven of them had been flooding. In part, that was because ours was a low-lying farm. But I wondered if it also had to do with changing weather patterns, which would mean we could expect more of the same, a long series of years that started with hope and faded into disappointment. That fear gnawed at me. I knew enough by then to see that farming was a game of betting for or against natural forces that were much bigger than we were. To nudge the odds in our favor, we needed to hedge against the things we couldn't control. The only surefire way to hedge against too much rain would be to move to the bone-dry Central Valley of California and get our rain on demand from irrigation canals. Otherwise, we could drain our fields.

Mark and I talked about drainage endlessly. We knew that our farm had been drained in the past because we had unearthed busted-up lengths of pipe-shaped ceramic tile, which farmers had used for drainage before the advent of plastic. Since the 1970s, farmers in our region had been using four-inch plastic tubing perforated with tiny slits. The excess water could seep into the slits and run through the tubes, which were buried four feet deep throughout a field, in regular lines that joined together into a main line that drew the water out. Any farmer whose economy was big enough to afford to drain the fields, did so. Drainage could increase production tenfold in fields where the soil was good but wet. It allowed farmers to plant earlier in the spring and get crops out in the fall. It took away those dreaded wet feet and helped plants thrive. Whenever I saw puddles in our fields or watched the leaves of plants turn yellow from flooding, I argued that we needed it.

Mark argued the other side. There were tillage techniques we could use and improvements we could make to mitigate flooding without the huge expense of adding drainage. Moreover, he had a visceral hatred of plastic and the diesel required to install it. But really, the reason we hadn't drained our fields was that we simply couldn't afford it. The cost of draining an acre of ground topped what it cost to buy an acre in our area, and we were already at our financial edge, paying the mortgage on what we owned, paying ordinary bills, meeting a growing payroll every two weeks. Sales were up, but at this rate, drainage was probably ten years away, and that was without setbacks. After a year like this, if bad luck continued, we wouldn't make it that long without going bankrupt.

Then one hot, muggy day in late summer, Mark walked into the ice cream shop in town. Everyone in the village ends up there on summer evenings, locals and tourists alike, spilling out onto the curb or walking along Main Street with a dripping sugar cone, past the library, down to the ferry dock or the rocky shore of the lake.

A woman approached Mark as he ordered. She said, “I’d like to talk with you alone sometime.” Mark felt himself tense up. This sort of request made him nervous. Most people in town had been so supportive of us and our farm, but there had been some friction too. Our first year, in the same ice cream store, the man at the counter had scooped our ice cream with melodramatic contempt. We had no idea why. It took inquiries to find out he was angry because we had special permits to shoot deer in our fields out of season. He was an avid hunter, and our permits might lower his chances for a buck. Another year, a neighboring farmer, who was formerly friendly, became blisteringly angry with us because we’d taken over his grazing lease on some land adjacent to our farm. Other people disliked the sight of our run-down buildings, which cluttered the view so close to our quaint little village. So when the woman made this vague request, Mark demurred. He wondered if she wanted to complain about the crummy office trailer we’d plunked down near the road, or the rusty horse equipment parked in a line next to the driveway.

The next week, he was walking alone down the farm driveway, a damp notebook in his hand, running through the list of things he needed to do before sleep. The day's heat was radiating up from the ground in thick waves, and he was sweating through a dirty shirt. The woman from the ice cream shop pulled up next to him in her car, leaned over, opened the passenger door. "Get in," she said. She had the air-conditioning on full blast, and he sank warily into the cool seat. She handed him a cold Coke. "You probably don't drink things like this," she said, grinning. And normally, he didn't, but he took it and drank it gratefully, the sweat from the cold can dripping mud onto his lap from his soil-covered hands. "I want to get right to it," she said. "I want to pay you to drain your fields. Twenty-five acres."

Mark looked at her, uncomprehending. He told me later he felt like the woman in the driveway, with her air-conditioning and soda, was playing the part of a genie, or some other plot device in the drama of our farm and our family, too perfect to be believable, a deus ex machina dropped from the hot sky into our sad field. "Why?" he asked.

"Well," she said simply, "I've been watching this place for the last eight years. Your fields are wet, and you need drainage." Mark kept his mouth shut for once, numbers clicking through his head. She exhaled and continued. "I've lived in this town for twenty years," she said. "I love this town. And your farm has made a lot of positive changes here. I want to see that continue. But you need to drain it."

I knew the changes she meant. They had something to do with Mark's incessancy. He was so determined to make this farm succeed. Rural America, family farms in America, had been struggling against a sense of defeat for so long. Our corner of New York was no exception. When the first wave of dairies went out of business in the 1980s, the feeling took root that there was no way to win. The hard-won skills that rural people possessed had lost their value. Then the rest of the world got high-speed Internet and a booming economy, and we were left behind. Alienation led to hopelessness led to torpor, and the young moved away or gave up. The older people saw that and lost hope themselves.

Mark might be irritating sometimes, he might even be delusional, but he was rich in hope. And hope has its own energy. His hope, and the clear vision of the farm that he projected, brought young people from far away who wanted to live here and work with us. Our farm was close to the village, and the best fields were right by the road. These strong and beautiful people were visible every day, working hard, impossible to miss. They rented houses in the village, fell in love with each other, got coffee in the café, went swimming at the town park. They brought young life to town. Soon they were starting their own farms, and there was talk about a food renaissance, the possibilities it held to draw new families to us, to better everything for everyone.

It doesn't take much, in a tiny town, to shift things. Our payroll that year was only $150,000, but a good part of it was spent in the village, which was enough to make a difference. But mostly, it was the visibility of the work. It heartened people to see. "How is it you have *seven* beautiful women working in your field?" one of our neighbors asked Mark. "Are they beautiful?" Mark asked, genuinely surprised. He's like that. For all our other conflicts, he has eyes only for the dirt and for me. He hadn't noticed, but everyone else had.

In the car, with the woman, Mark chose his next words carefully. "What do you need in exchange for your generosity?" Mark isn't like most people, like I am. He has no manners, really, no compulsion to refuse an extravagant gift. I probably would have tried to thank her for the thought and escaped the car as fast as possible. It never would have occurred to Mark to refuse, if the life of the farm was at stake. But he did see money as energy that needed to be balanced. He wanted to understand what would need to flow back from us to her.

"Nothing," she said. "I don't want anything for this. It's not a loan, it's a gift. I don't want to be thanked. I don't even want to be invited to dinner.

I'd prefer anonymity. You have a wet farm, and you need drainage. Drainage will help you grow. I'll send you a check next week." And she did.

Mark is less interested in money for money's sake than anyone I've ever met. He would *rather* drive an old car, if he must drive a car at all—owning a car, as he often reminded me, was a concession to the life of semi-normalcy that he'd agreed to as part of the bargain of marrying me. Before we met, he biked when he needed to get somewhere, or hitchhiked. He'd *rather* get his clothes secondhand and wear them until they literally fall off of him. If the woman had handed him a diamond, or any priceless beautiful impractical thing, he would have stared at it and handed it back to her. But what she handed him was a renewed ability to pull things out of the earth. And the energy of her generosity cleared away any remaining doubt that our farm would grow.

There is nothing that animates Mark more than taking an idea out of the air and making it reality. If he has the resources to do it, it happens like lightning. An hour later, he was on the phone, calling around to fellow farmers, to the extension office, to excavators, getting leads on who lays the best tile drainage. Within the week, he had zeroed in on a company, had a date to begin. The same day the promised check arrived, the Barnes family unloaded the drainage plow at our farm. Mark and I had decided to use the gift on the fields we called Small Joy and Large Joy, two low-lying fields east of the farmhouse that had beautiful sandy-loam soil but a problem with flooding.

I took Jane to watch the last few lines of tile go into the ground. She had just turned four. I held her hand and explained that this was a momentous occasion that she should try to remember. We met Milton Barnes, the family patriarch, eighty-five years old, who had been laying tile back when it was actually made of ceramic tiles. He was at the bottom of a six-foot ditch, wielding a shovel, while his sons drove the backhoe and the drainage plow. It was good to know we were in such experienced hands. The drainage plow itself looked like an ancient monolith attached to the bulldozer's insectlike tail. It dug a narrow trench five feet deep and fed the plastic tube to the bottom, where it would drain excess water to the edge of the field. The plow was blunt, colossal, and we walked behind it, watching the brute force of diesel at work, the plow running through all that earth as though it were made of warm butter. We stopped, dangled our legs in the cut, and held handfuls of the dark, loamy topsoil and the light, sandy subsoil. There were no rocks to be seen. The two fields were one now, and needed a new name. When the surface had been plowed and harrowed smooth, we seeded it to rye, as a cover crop, and rechristened it Superjoy.

Just after the drainage was in place and the new field was seeded, we began to hear dark words on the weather radio. A hurricane was building off the East Coast, tracking toward us, bringing heavy rains and high winds. They were calling it Irene.

We did what we could to prepare. Mark secured the metal tops of the broiler chicken coops with old tires, while I moved the laying hens to higher ground. The tomato plants were heavy with fruit that would be destroyed in a big storm, so we harvested hundreds of pounds and stored them in the pavilion. The night before the storm hit, Mark and I put the kids to bed and walked the farm, looking for loose objects. We nailed the doors to the barn lofts closed, and weighted the tarps over the grain bins with logs and heavy rocks. Then we climbed to the top of the pole barn with hammers, crawled along the roof, and nailed down all the rattling sheets of metal. From up high, we watched the sun set. The wind was picking up. There were no clouds gathered yet, but the sky over the sugarbush hill to our west was stained a dark and ominous red.

The storm came that night and lasted all through the next day and into the next night. The kids and I watched from inside. The rain fell in heavy waves and poured off the roof in sheets. The wind raged, pressing against the windows and doors. In the middle of it, Racey appeared, and she and Mark made their way to the field where the broiler chickens were pastured. The field was flooding, already four inches deep in the spot where the coops were parked. The birds were huddled together in piles, three birds deep, soaked, and trying to stay warm. Mark and Racey tossed them into a single layer to try to save the ones beneath, but some of them were already dead from a combination of chilling and suffocating, and no sooner were they liberated than the live ones began to bunch together again. As branches crashed in the hedgerow, Mark and Racey wrestled bales of hay into place and pulled the coops up onto them, slightly uphill, so that the chickens could get out of the water and stay warm.

When the storm finally blew itself out, we'd gotten ten inches of rain. The power was gone, and the sweet corn was flattened and tousled like it had bedhead. But the tile drainage was working, drawing hundreds of gallons of warm rainwater out of the field every minute. Within two days, instead of deep puddles, we could see the lines of green rye beginning to germinate in the good soil.

We'd been very lucky. Aside from the dead

chickens and a few fences that needed repair, we were okay. Over the next week, we got messages and photos from farming friends all over the East Coast whose good bottomland had been swallowed by floodwaters, their homes and buildings damaged. Still, that storm broke me. *What is the use of this work when it can be taken away so easily?* I thought. *Why should I risk loving something this much if it's so hard to hold on to?* And just a little deeper: *Why should I stay in a marriage that's hard? Is this really a good place to raise children?* The storm brought out all those water emotions in me, and I felt soaked in fear too big to speak about, at least directly.

Instead, I got remote and prickly. Mark and I chafed at each other. Rather than talk about the important things, we argued about the little things. He wanted to add a new greenhouse. I thought we couldn't afford it. He wanted to open a line of credit so we'd have easier access to capital when we needed it. I didn't want to risk it. The imbalance between what the farm got and what the family got was eating at me. Winter was coming again, and the dim house with its poi-colored walls felt like it was getting smaller around us. I fell asleep listening to the sink leaking in the bathroom, as it had done for five years, all that time the farm's demands more important than the hour it would take to fix the faucet.

• • •

Racey and Nathan were leaving. She finally told me herself. She'd taken a job in Africa that started in a few weeks, and when she returned, she and Nathan were going to work at the farm down the road while they looked for their own land, to start a farm a lot like ours. It hurt that I had heard it first as whispers and rumors. It hurt to be abandoned, and for another farm, no less. And it hurt that they were looking for land nearby, with a plan to directly compete with us for the very small market of people in our town. It wasn't the first time this had happened. Another couple we'd hired had started a farm just down the road that was also powered by horses and offered the same kind of full-diet membership we did—in fact, that was where Racey and Nathan were heading. Tobias and Blaine were looking for land in the neighborhood too, planning to start a farm that offered grass-fed meats. But it hurt more with Racey because she was my friend. I'd known it was risky to let myself become too close to her, but we had a lot in common—similar education, worldview, and interests, similar love for travel and for farming.

Mark saw things completely differently, as usual. He thought the world needed a lot more small, diversified farms, and if we could help spawn them here, in the place we loved, then that was a good thing—a great thing. Yes, there would be more businesses producing for the same market, but we could choose to see that as competition or look for the ways in which we could collaborate. "Imagine if we have enough farms here in ten years for someone to reopen the farm supply store!" he said. "We wouldn't have to drive two hours to get parts when the mower breaks. Imagine if we could buy our supplies cooperatively. We could have a real functioning small-farm economy here for the first time in a hundred years. It'd be good for everyone." I was skeptical. That didn't solve the problem of sales. I couldn't shake the feeling that doom was right around the corner.

Jenny arrived to take Racey's place. She'd just shown up in our dusty driveway one day, in a car with her native Ohio plates, carrying all her possessions. She had sent us a résumé and left some messages over the summer, when we'd been too overwhelmed to follow up. When she didn't get a response, she just got in the car and drove east, then announced that she'd be working for us. She was in her mid-twenties, tall and strong, with a wide and constant smile and thick brown hair. She'd gotten an undergraduate degree in urban planning, then figured out she was interested in food. She'd worked in the kitchen at Chez Panisse, then spent a year working at a horse-powered farm in California, and she knew she was onto something, so she came to work for us. Without hesitating, she jumped in and began to help.

The first Sunday she was with us unfolded in the usual mix of routine plus surprises. I was in the house with the kids, and Mark was in the barn. One of the dairy cows, Camden, calved that day, a sweet little heifer. Mark carried the calf the whole mile up from the dry cow pasture on his shoulders, got her settled in the calf nursery, and tended to Camden's swollen udder. The girls and I joined him and fed the calf her first bottle. Then we cleaned the house, made lunch, washed the dishes, swept the floor. Jane drew a picture of me dressed up as a clown for Halloween: purple shirt, green pants, wild hair, and juggling balls. I asked her if she might want to be an artist when she grew up.

"No," she said.

"What about a farmer?"

"Nope," she said, without looking up from her crayons. "Way too much work." Then a long pause. "Mom? Do *artists* get to sleep?"

"Yep, they sleep."

"Okay, then, I *might* want to be an artist."

After their naps, the girls and I took a plate of sugar cookies over to Ronnie and Don. Milking had already started by the time we got back, and I knew Mark would need help with the new calf, so

I pulled up to the barn. Jenny was helping Mark milk. I told her I'd get a quick bite of food into the girls, grab our boots, and be right back out. But as I was getting Miranda into her high chair, Jenny came into the house, looking for a quart of vegetable oil. Mark had sent her to tell me that Juniper—daughter of June, a granddaughter of good old Delia—was down, and he suspected she had bloat. Bloat? I was skeptical. On our farm, bloat is usually a spring problem, not a fall one. It happens when ruminants graze lush pasture that is rich in legumes, like clover or peas. When any sort of feed hits a cow's rumen—the first stomach—it mixes into a very rich broth of microorganisms and begins to ferment. Eventually, it forms a cud, which the cow will regurgitate, chew again, and swallow, to finish digesting in her next stomachs—reticulum, omasum, abomasum. The rumen does the grunt work. It can hold nearly fifty gallons of material. The dense mix of microorganisms in the rumen breaks down the cell walls of plants. One of the by-products of all that biological activity is gas. Normally, the gas bubbles are released through the mouth. But in bloat conditions, the legumes cause the rumen contents to get frothy, like beer from a keg just before it's kicked. The gases from fermentation are trapped in the froth. Because the gas can't be released, pressure builds inside the rumen. It can happen very quickly. The left side of the cow's body puffs up and tightens like a drum. If it progresses, the rumen presses against the other organs, crowds the diaphragm and lungs, and the cow can't breathe and dies a painful and miserable death. Could it be bloat? True, all the rain we'd had made for some gorgeous lush grazing, and they were on a field rich in clover. It was possible. In any case, the treatment—a drench of vegetable oil—wouldn't hurt her. Oil would hit the foam in the rumen and cause it to settle so the gas could be released. If that didn't work, the emergency lifesaving alternative would be stabbing a hole into the swollen balloon of rumen with a trocar—a gruesome operation that would make it deflate all at once and bring relief but also a chance of infection.

I filled a beer bottle with oil and gave it to Jenny. A few minutes later, she came back, asking for another. I quickly got the girls dressed and in the double stroller and walked through the pitch dark to the barn.

Juniper was clearly in distress, lying on her side, left flank distended, looking like she wanted to die. Mark and Jenny had been trying to haul her up onto her feet, and she wasn't having it. I joined them, and we were able to get her up and into a horse stall before she went down again. At least the angle of the bedding made her front higher than her rear, and put her mouth above her rumen, which could help the gas come out. Mark reached into her mouth with his hand and tied a stick between her teeth. She worked at it with her tongue. This was supposed to help encourage her to burp and release the gas. After a minute, she let out a tremendous belch. And another. And three smaller ones. Jenny watched, mute. Juniper rose to her feet. Cows are strikingly unexpressive about pain—they are prey animals, and in the long game of evolution, no good came to a prey animal that revealed its wounds or weaknesses—but with each burp, you could see a shift in her energy, back toward wanting to live.

By then, it was past suppertime, and we hadn't even started to milk. Jenny could help Mark finish milking and feed calves. I'd get the kids fed and put to bed and then bring some dinner to the barn.

Back in the kitchen, things didn't look very promising. On weekends, I made a point of using up leftovers, and by Sunday evenings, we were sometimes down to half a jar of yogurt and a collection of what Jane derisively called "cold scraps." The kids were hungry and chilled from the barn, and Mark and Jenny would need more than cold scraps too. I found six cooked potatoes, half a pork chop, an onion, and a mixed bundle of fresh herbs. I needed something that would be fast, fill five bellies, taste nourishing, transport easily, and be eaten with hands. This was a situation that called for fritters. I pushed the potatoes through a food mill, grated the onion, and cut the pork chop into bits, then added eggs and a handful of flour, plus salt and pepper and all of the herbs, chopped fine. I heated an obscenely large spoonful of lard in the big cast-iron pan and fried the mixture in blobs that turned golden brown, crisp on the outside, fluffy and tender on the inside, shiny with hot lard. I left the kids at the table, a fritter in each fist, and ran out to the barn with a steaming plate for Jenny and Mark and a jar of sour cream.

Juniper was already back in her stanchion, eating dry hay, as though nothing had happened. Mark sloshed some of the fresh warm milk into mason jars, and before I went back to the kids, I stood in the aisle of the barn, drinking warm milk and eating hot fritters slathered in sour cream. I checked Jenny's face. Was it too much for her? This wasn't Chez Panisse. But she was smiling widely in that way that let me know she got it. That she was good with this crazy, dirty, brutal life.

CHAPTER 12

The farm changed into her early-autumn dress, dark green with red and orange accents. I lay in bed before dawn, listening to the geese honking their way south. The weather was serene that fall, all its rage already spent. The honeybees plundered the goldenrod, returning to the hives with their saddlebags full of gold dust. We shifted the solar panels to their fall positions to catch the fading light. It felt bittersweet to watch it go. The tilt of the world tells us there's an end to everything.

No more planting but not much weeding either; weeds have to obey the light just as the crops do. We made a last few acres of hay, using the horses on the sickle-bar mower to cut it, because every bale that went into the mow was another small piece of security against the coming winter. The cover crops of rye and vetch grew into a fertile green blanket over the bare ground on the newly drained field. The leaves on the pumpkins and winter squash withered and died back, exposing their finished work: orange globes, beige blobs, and green-striped yellow zeppelins. On cold mornings, instead of lighting the woodstove, the girls and I filled the oven with pumpkins, roasted them, then scooped out the flesh to freeze for a winter's worth of pumpkin soup, pumpkin pie, pumpkin cake. The tomato plants slowly succumbed to early blight, death creeping up from the ground into their leaves and vines, which were still heavy with cracked green fruit. The celeriac grew larger, and so did the potatoes, carrots, and beets. The sweet corn was tattered by the hurricane, but the stalks were vertical again, lifted by the sun.

At dinnertime, I put a pot of water on the stove to boil, and took the girls to the field to pick sweet corn. The leaves were raised like a church full of believers waving hands in the air to catch the spirit.

Jane was in a tutu and tiara phase, and had

spent most of the summer in a pair of sparkly red shoes, the glitter now worn off at the toe and embedded with soil. The ensemble made it easier to spot her even when she ran far off between the shady rows. I heard her break off a low ear, shuck it, and munch it raw, a farm-kid appetizer. Miranda was beginning to walk, and she pulled herself up on a sturdy stalk. I yelled to Jane to hunt for *huitlacoche*, the lumpy fungus that is a delicacy in Mexico. It was one of our favorite treats. She shouted back when she found an ear of it, and it was a good one, not yet blown out in black spores but bulging with firm lobes. We put it carefully in our bag and then chose our corn for dinner. I craved ears that were as mature as the season felt by then, with hulking kernels that I could sink my teeth into, a complex corn character, not the extra-sugary kind with kernels that pop off into your mouth when you bite. On the way home, we stopped at the raspberry patch. The storm had melted the oldest berries into a fermented mush, but a fresh wave of them had ripened. We had forgotten to bring a berry box with us, so we pulled off Miranda's shirt, made a pocket of it, and filled it with a quart of berries.

The water was boiling when we got back to the house with our loot. Into the steam went a dozen pieces of corn. Fully half of them were for me. Mark still could not get over the fact that his bantamweight wife could eat six ears at a sitting. Jane cut the *huitlacoche* from its host cob, and we sautéed it with butter, salt, and green onion, and a sprinkle of chopped cilantro. Mark came in from the field and lay flat in the middle of the floor for a ten-minute nap, which he fell into, as usual, within seconds of lying down, while Miranda crawled over him. Then he came to the table and we feasted on corn, some of it rubbed with butter and sprinkled with salt, some slathered with homemade mayonnaise, and some dusted with hot chili and spritzed with lime.

For dessert, I pulled a pint of our heavy cream from the refrigerator. The dairy cows were thriving on the late-summer grass and clover, and the cream was the golden color of a milkmaid's dreams, the texture of cake batter. We spooned it into bowls that we topped with the raspberries, which I had cooked with a vanilla bean to a vibrant bubbly sauce, and a swirl of buckwheat honey from our hive. We ate it all and licked the bowls clean.

Nobody went hungry that year. The generosity of what the farm provided for us and for our members, the abundance and variety and sheer deliciousness of our food even in the worst year we could have imagined, should have buoyed me. Instead, I couldn't shake free of the fear that what we had endured could happen again. When

Mark saw that, he wanted to erase it. Maybe his own happiness was too fragile to risk being dulled by my doubts.

One evening we sat at the kitchen table, going over the books and planning the work for the week. I could feel the tension in my shoulders and jaw. The numbers looked scary, and the list of tasks was both daunting and urgent. As Mark talked on, enthusiastic, I stopped listening, became fixated on the steady drip of the drain under the kitchen sink, which was leaking into a dog dish, as it had been for months, old dishwater collecting there in a viscous gray soup. *I need to get out of here,* I thought. I hadn't been off the farm in days, maybe weeks. Actually, I couldn't remember the last time I'd left the farm. No wonder I was feeling a little crazy. A day in town would do us good.

I looked up and interrupted him. "I'm going to take the kids to Plattsburgh tomorrow," I said. "I'll pick up the parts we need to fix the drain." Plattsburgh was a forty-five-minute drive north, the city we went to when we had to buy supplies, see the dentist or the pediatrician. I'd had a sudden vision of the three of us, whizzing through the aisles of the big-box hardware store, the baby in the seat of the shopping cart, the toddler holding on in front, as I bought the mysterious parts. Maybe I'd stop in the paint aisle too, for something to spruce up the walls of the bathroom. Maybe, afterward, I'd take them out for hamburgers.

I saw a ripple of displeasure cross Mark's face. "You won't know what to get. It's not so simple. It requires some research. It's not important right now. And anyway, if you have *that* much time, we could use your help in the field. It's harvest day, and we're shorthanded."

My fear that he loved the farm more than he loved us was counterbalanced by his fear that I didn't love the farm *enough*. That my commitment wasn't sufficiently deep to keep it alive now that we had children. That if I strayed too far from the farm's gravity—that hard constant pull of responsibility and work—I'd be swept beyond its reach. Every time I wanted to leave, even for an afternoon, he came up with a list of reasons I should stay. It wasn't that I couldn't go, I knew, but that the fight would be too tiresome, and there would be smoldering resentment in its wake. "Fine," I said, and walked upstairs to bed.

When I close my eyes and picture that period of our marriage, the image that I see is this: a late-summer patch of ground. The soil is rich and well-watered. The sun is bright. There are good crops coming ripe, but the rows are overrun by our worst weed, galinsoga. The curse of galinsoga is its seeming insignificance. It grows low to the ground, without the powerful grandeur of the redroot pigweed or the spiky menace of a thistle. Its tiny white and yellow flowers are delicate, pretty, like daisies grown for a mouse. But get close and you'll see that though it has barely bloomed, the white and yellow flower has already gone to seed. Closer, and you'll encounter its pungent stinkbug stink.

The tiny seeds of our disagreements were like that. So were the resentments I'd harbored when he was injured, as well as every time he did or said something that made me feel we were less important to him than the farm. They were insignificant until they multiplied, each one dropping a new generation of wrath. They choked out the beauty until it was hard to see.

In the wake of the drainage, the purgative of Irene—Mark was lit even brighter than before with enthusiasm for farming. His natural reaction to stress was to throw himself toward it and turn up the amperage of his default optimism. My instinct was to retreat, back to a time before we had to deal with so many zeroes on our expenses, or employees who were often disgruntled. Despite the bills on the table, Mark wanted to make some major, expensive improvements. We needed to reconfigure the well, to make it more reliable, and as long as we were going to fix that, he wanted to find the money to invest in an underground corridor that would carry beefed-up electrical wires, water pipes, and telecommunications infrastructure all the way from the farmyard to the pavilion. It would cost ten thousand dollars and do nothing for us in the short run. But what we had in place was antique, with limited capacity. It would not support a farm any bigger than we already were. He made a convincing argument, based on how smart it would be in the long run, and I agreed, but not without a shudder of dread.

Mark was so enlivened by work and what he wanted to accomplish that he sometimes woke up at midnight that fall and couldn't wait to start. When I got out of bed at five, I'd find him cooking breakfast in the kitchen, humming, having pulled the stakes from the spent tomato patch by headlamp or completely reorganized the machine shop, then run the four miles around the farm for the joy of it.

It doesn't feel great to be the shade on anyone's lamp. I didn't want to be so down. Maybe I just needed to get out there again and work. Now that Miranda was walking, my shoulder was starting to heal. In mid-November, I took over the evening animal chores on weekends. Mark would watch the girls. On my first Saturday, I kissed them goodbye, pulled on my warm overalls, mittens, and hat with furry earflaps, and walked out to feed the cows.

The late-afternoon sun was weak, coming through dense clouds. The wind and the damp air made for a chill that felt colder than the bright deep-cold days of midwinter. In the driveway, the mud was half-frozen, the texture of pottery clay. But it was good to be outside, moving, and free to walk at my own pace without the weight of a baby on me. Jet was with me, happy we were at work together, pacing along at my heel.

The dairy cows were pastured just behind the East Barn. They would need thirteen bales of hay, which was stored in the loft. The square bales we'd made were poor quality, but at least they were plentiful, good enough to fill the cows' bellies while they grazed the end of the stockpiled grass. The last of them, put up in late September, had filled the loft completely; the final wagonload had to be wedged tightly against the slanting boards of the roof, leaving only a small crawl space down the center.

The bales covered the southern door to the loft so that the only way in was through a hole in the ceiling inside the barn. There were a few boards tacked between the barn's beams to make a rickety ladder. I climbed them to find that the bales above my head were stacked nearly flush with the hole, only two inches of wood around the edge. When the ladder ended, I had to get a toe on the loft floor and climb straight up the bales, holding on to their strings, my body hanging over the dark abyss of the hole. I felt a flutter in my stomach.

There was no light in the barn; I reached up to turn on my headlamp. The battery was dying, and the light that the lamp gave was barely enough to see the bales in front of my face. There was a metal cable halfway up the stack, which tied the beams of the barn together. I grabbed the cable and a bale string and pulled myself up, then planted a foot on the cable and felt for the top of the bales. There was not enough space to stand, so I crawled away from the hole, on top of the bales, until I reached the window at the northern gable end. I threw thirteen bales out of it, into the pasture below, watching them crash onto the grass. Next I needed to climb back down the ladder, toss the bales into the feed bunk, then open them with my knife and spread them out so all the cows could eat.

By the time I crawled back to the exit hole, my headlamp had gone out completely. There was still light in the sky outside, but away from the window, the dark in the loft was complete, no hint of shapes or shadows in front of my eyes. I was a little frightened, thinking of that cool vertical expanse of dark I had to navigate, and the fear made me hurry. I found the edge of the bales with my hand, held on to the strings, and sent my feet over the side, feeling for purchase. Then I grabbed for the cable with my mitten-clad hands. The idea was to hold on to the cable, keep my feet against the bales, and slowly inch my hands down until my toes found the two inches of floor. Then I could creep around, holding the bale strings, until I got to the ladder. But I was too hasty, and my hands were too cold, the mittens too slippery. My hands slipped off the cable, and from the top of the stack of bales, I fell.

I had time to think, *This is really happening,* and feel the rush of black air as I went down through the hole, and then a wordless flash, a glimpse in my mind of Mark and the kids, before I whomped into a pile of loose hay on the floor of the barn. I had fallen fifteen feet. By some miracle, my flailing arms had snagged a loop of hose suspended from the barn's ceiling on the way down, and I'd caught it against my body. That knocked the wind out of me and bruised my ribs but also slowed my fall. Thanks to the hay and the way that I landed, there was nothing else wrong with me.

I lay still, feeling the wave of adrenaline crest and begin to abate, while Jet licked at my face, which was suddenly wet with snot and tears. It had been a long series of painful separations, beginning the year before with that irreversible snip of rubbery blue-white cord that had connected Miranda to me. There was the separation of me from the work of the farm, of Mark from me, of the farmers from the house, and me from Racey, the separation between workers and friends. There was the long season of disastrous weather that seemed to separate us from a secure future and raised doubts about whether I wanted our story to continue. There was the looming stretch of midlife ahead, of youth behind. The fall from the loft was an abrupt end to it. It snapped something between me and the past.

There was no use holding on to the way Mark and I used to be together, when we were a couple and not a family. Did we work in our new shape? The family and the farm propped each other up even as they wore each other down. There would be no farm without the marriage, and I suspected there would be no marriage without the farm.

The farm had to grow and change in order to live, and we'd have to grow and change along with it if we wanted to keep it. There was no guarantee we'd be okay. There was no guarantee the farm would survive. The risks would always be there. I could grieve the loss of what had been and reach for what was coming. I'd have to, if I wanted to stay.

Because there was also a big question in that long split second of dark descent. *Do you want this?* This life, this farm, this man, this marriage, this struggle to make it all work? And an answer. *I do*.

For the next few weeks, I could feel bruises on my ribs when I laughed or breathed deeply, which reminded me to be grateful that I was still doing both of those things. I kept doing my chores and picked up a milking shift, which helped put me back in touch with the daily work of the farm, providing exercise, purpose, and fresh air. And I called Candace, a therapist who lived in the village. She agreed to sit down with Mark and me, to help us try to put ourselves back together.

The holidays arrived. We hitched Jay and Jack to the hay wagon, gathered all the farmers, and went caroling through the village. We had my great-grandparents' sleigh bells for the horses, and fifteen of us on board, trotting through the cold night. A flask of cheer went around among us, and the sound of our singing got louder and more off-key. Mark drove the horses, Miranda snuggled into my arms, and Jane leaned into my shoulder.

The week of Christmas, there were chores to do, cows to be milked, and most of our farmers were gone to see their families, so Mark was in the barn or the shop, or rushing between groups of animals, keeping everyone fed, watered, and bedded. Two days before Christmas, I strapped Miranda into the backpack and put Jane in her snowsuit and trudged out through the soft snow, stopping at the shop for a saw, in search of a Christmas tree. We found a spindly spruce, good enough, in the no-man's-land north of the farm road. The half-grown pigs were pastured there, and as we stepped carefully over their low electric fence, they snorted at us, then ran off together like a school of pink fish. I cut the tree down near the ground and hauled it back to the house, stabbing the baby in the process, and prickling Jane, who was already crying, because she had gotten snow in her boot.

The trunk was too small for our tree stand, and it refused to stay vertical, so I stood on a chair, screwed a hook into the crumbling plaster ceiling, and tied the tree precariously up with a piece of fishing line. It dangled there, ridiculous. The project had taken most of the afternoon, and there were more tears. But at last, it was done. While the girls reverently fingered through the box of ornaments, I put on some Christmas music and made popcorn to string. Then I stoked the woodstove and poured myself a little glass of port. As the heat filled the house, I sniffed at the air, detected something alarming under the smell of popcorn, and slowly zoned in on the source. A dark and disconcerting stink was coming off the boughs of the Christmas tree. I got closer and found I had dragged the tree through the pigs' half-frozen manure, and the green needles had deftly picked it up and held on to it through all the tugging and righting of the tree, until the heat from the stove released it.

On the last day of the year, the girls and I made paper boats to hold our wins and losses. On one side of my boat, I wrote down blights and injury and financial stress, all the marital conflict, the fatigue, and the endless god-awful rain. On the other, I wrote down the good things: two healthy children who made my heart squeeze with love every time I saw them. Our expanded dairy, a mighty team of farmers, twenty-eight acres of drained ground with gorgeous soil, the strong horses in big hitches, pulling the equipment. And one last word, in all caps: FOOD. Even that year, the food had delighted and nourished us.

At sunset, we took our boats to the lake. It was ten degrees outside. Miranda was sixteen months old, a twenty-pound live weight with a strong will. Sometimes at the end of the day, I just didn't have the strength to wrestle with her. I decided to skip the frustration of mittens, snowsuits, hats, double socks, and warm boots—all of which would go on with tears and complaints and then probably have to come off again before we were out the door, because one of them would need to pee. I warmed up the car and left all the gear in a heap by the door. I put their jackets on, hoods up, and tucked a random selection of mittens into my pockets. On the way down the driveway, we ran into Mark, between chores, and he jumped into the car with us.

We got to the park, and I slid the car down the ice-covered slope to the lake. I didn't have snow tires, but there was one bare strip of pavement showing, enough to stick a tire to for the climb back up. The lake lapped at the town dock, each wave adding a layer to the thick coating of ice on the windward side. The ferry was coming in; the water looked more substantial than usual in its few degrees above freezing, as though it were already thinking of going solid. I left the car running and popped Miranda out of her car seat and carried her over the snowbank onto the dock with her scribbled-over paper boat clutched in her bare hand. Mark brought Jane on his hip. She had made a tiny paper bowl for a boat, the size of a pea, just big enough, she said, to hold her wishes. He set her down at the edge of the dock, and she leaned forward over the water to throw it. I grabbed the back of her jacket so she wouldn't fall in, and she leaned even farther, then launched. I tossed Miranda's boat for her, and then I threw my own. It landed sideways, rode the waves for a few moments, and sank out of view. Then we turned our backs on that difficult year, and we all walked away, together.

PART 3

REGROWTH

CHAPTER 13

Nothing really gets better all at once except in books. The streak of bad weather lasted another two years. Without drainage, we wouldn't have made it to our tenth anniversary. With drainage, there were still plenty of moments of uncertainty and despair.

We scrambled, financially, to make up for the setbacks of weather. We lost some members to the new farms around us. We took on more debt to pay for the grain and hay we couldn't grow, the new well, and more drainage to the fields. We needed to increase our sales in order to service the new debt, and the local market had gone flat, so we started delivering our year-round share to New York City every week. That changed our potential market from the few thousand people who lived close enough to get to us, to the many millions of people in and around the city. Sales grew rapidly, but the growth required a huge input of labor and infrastructure. Mark and I were both aware, as we lifted this new heavy project off the ground and tried to steady it, that the act of creating something new is easier when you're younger, and we weren't getting any younger. We wondered out loud how much more heavy lifting we had in us, if the farm should require it.

We got remarkably lucky. The wet years, when we were short on hay and grain, were followed by mild winters. Without the caloric drain of deep cold, the animals needed less feed, which saved us from having to buy too much. My first book brought in an unexpected royalty check just in the nick of time. We squeaked by, barely. One summer after all the ready cash was spent, we needed to buy a truck to replace the old Honda Civic, which had finally and decisively died. The Barnes family was on the farm that week, putting in more drainage on credit. As they were setting the lines in Pine Field, the drainage plow snagged some ancient copper wire. Whatever it used to service was long defunct. But it came snaking out of the soil behind the plow, hundreds of feet, buried treasure. We unearthed just enough of that valuable stuff to make an even trade with our friend Jason, who took the copper for scrap in exchange for a decent old Ford F150.

Mark's lights burned out two more times. The following spring, he injured his back again, not planting but in true hippie style: playing Ultimate Frisbee with the farm crew. That injury was at least as excruciating as the previous spring's. He was down for five weeks. The year after that, toward the end of winter, he took our whole farm crew skiing, to thank them for all their hard work. On the first foggy run of the day, just off the chairlift, he crossed his tips and broke his leg in such spectacular fashion that he required a four-hour surgery, a selection of plates and screws, and a big fat bone graft; he wasn't back to normal until August.

Both times, there was that same separation between us, and I felt the same lonely fear of failure and loss. Both times, we teetered. And both times, we pulled back together. Why? It helped to have been through it before and come out the other side. And because we somehow fit each other's unusual emotional topography. As my sister had told me, I could probably never love a normal person. Maybe a normal person could never love me. Once I was rifling through the chest freezer in the basement, looking for green vegetables. "Find any?" Mark yelled from the top of the stairs. "Nope," I yelled back, "just loads of tomatoes, some cow colostrum, and a bag of raspberries. Oh, wait, those aren't raspberries. That's Miranda's placenta." We'd buried Jane's just after she was born, and planted a sour cherry tree on top of it, but hadn't gotten around to burying Miranda's before the ground froze. "You should be very glad you're not in the dating pool," Mark called as I stuffed it back into the freezer.

Most of all, we stayed together because we had between us some important things that we had created, that it was our duty and privilege to nurture—two small children and one big farm. Even a radical farm imparts conservative values. It does so without you really knowing it. In the same way the military makes you a person who stands up straight, a farm will give you grit and perseverance.

Another big reason we made it through those hard stretches was Candace, our therapist who lived just a mile away, in the village. She might have been the only shrink in the world who could properly handle us both. I thought of her as part shrink, part coach, part shaman, which fit us exactly. She came bustling into our shabby house for weekly meetings, never minding that she had to walk past a broody hen recovering in a cage in the mudroom, or a calf's head that Jet had dragged out of the compost pile and left at the door, ghoulishly, faceup. Candace was always exuberant, sometimes singing. She brought a mix of intuition and empathy and complete acceptance along with her useful psychological training and many years of experience. She walked us in baby steps through the hardest parts of those years, gave us tiny assignments that we could actually manage, sometimes as small as to just look at each other for five minutes. She armed us with the tools we needed to live together in peace and, eventually, to fall in love again, in a different way, which allowed us to see each other more fully, to admire each other's good parts, and to tolerate each other's faults.

I got some feedback from Candace that helped me see where I was, with the clarity of perspective that I didn't yet have. Part of the difficulty, those years, was feeling like I was doing it wrong somehow. I believed, deep down, that raising kids should have been easier than it was. My own mother had made it look not only easy but almost always extremely fun, and she did it without five hundred acres of space for us children to roam. Of course, Candace pointed out, my mother also did it without a farm to run or a job outside the home. I'd somehow fused the sweet memory of my stay-at-home non-farming mother with the vision of the twenty-year-old mother who had visited us and farmed all day with a baby on her back, and I really believed I should be able to do both things gracefully. If I couldn't, if it wasn't productive, fun, and easy all at the same time, then I must be failing. "Nope," Candace corrected me. "You're in the trenches of young childhood right now, and it's really fucking hard." The thing about a trench is that you can't see much outside it. Candace gave me a very specific, very unshrinky forecast, which I grabbed and held tightly on to in the dark times. "It all gets a lot easier," she said, "when your youngest turns five."

I still suspected that more experienced mothers figured out better systems, a different set of rules. If I was struggling, how could families manage with more than two kids? I'd heard people in large families say that it gets easier after the fourth because the older ones can help. I was walking down the farm road with an Amish grandmother a few years ago, chattering about children. "How many grandchildren do you have?" I asked. "Oh, almost a hundred, I guess. I've lost the exact count." At first, I thought she was pulling my leg in the dry Amish-humor way, but then I did the math. She had fourteen grown children, and they had their own large families. A hundred was entirely plausible. If anyone knew the secret system that made raising children a snap, it would be this lady.

"I've heard that it gets easier," I said, repeating my grain of wisdom, "after the fourth child comes along. Is that true?"

I saw her mouth twitch around the corner of her black bonnet. "Easier? Well, no, I'd say it gets a little harder every single time," she said evenly.

But it did get easier, bit by bit, as the girls got older. Mark and I learned that we needed to keep working to define my role on the farm so I was a part of it but could also accommodate the children and their needs. What I could handle changed as they grew, but it always involved adjusting my expectations of what a productive day looked like, and prioritizing patience over checking items off the to-do list.

We learned to call in help when we needed it. Ronnie swooped in at crucial moments to scoop up a fussy child or a toddler on the verge of a tantrum. Sometimes when I was needed in the field, she would come over to watch the girls, and I'd return in the evening to find her canning tomatoes for us in the kitchen or cleaning onions in the basement, the children contentedly asleep, the laundry done, and the house as neat as it ever got. On Friday nights, she would contribute something delicious for Team Dinner, and she and her husband, Don, would join the fray, elder statesmen among the wild and disheveled youth. Later, it became routine for our girls to go once a week to their house, where they got to be regular American children, with television, central heat, and wall-to-wall carpeting that Jane would describe to me later in hushed and reverent tones.

Another pillar of love and stability was Barbara. As the young farmers who worked for us came and went, an ever shifting cast, Barbara remained. She gravitated to the dairy, where she helped milk and make cheese. Her cooking was legendary. She could make something delicious and civilized out of any given ingredient—including a ragout made from gizzards and chicken hearts that became a popular farmhouse standard. Her apple tart, which she had learned to make as a girl, working at her uncle's restaurant in Adelboden, would appear regularly on our table. In her spare time, she liked to knit and weave. Like Ronnie, she had a sixth sense for when our girls needed a break from the farm, and would appear at the door with an offer to take them out on an adventure or over to her house for tea. The way she nurtured our children was different from the way Ronnie nurtured them; she gave them a pragmatic sort of Swiss love that was a perfect counterpoint to Ronnie's softness. Because of Barbara, our girls could knit before they could read, and they learned to spin yarn by hand from a raw fleece.

When the girls were two and five, Ronnie gave them their own little kitchen knives, sized for children's small hands, with rounded tips. The handles were shaped like dogs' heads, the cutting edges rolled like butter knives, safe for playing but not that useful for actual cutting. Mark watched the girls mush at a bunch of chives while standing on their stools at the kitchen table, then silently took the knives out of their hands, walked to the shop, and put them to the grindstone. When the knives came back, the edges were chef-sharp, though he'd stopped short of honing pointy tips on the ends. He stood over them, took their little hands in his big ones, showed them how to hold a knife, how to keep their fingers tucked out of harm's way. From then on, they knew how knives worked, and they spent hours absorbed in small, real jobs with me in the kitchen. They hunked carrots into chunks, chopped herbs, peeled and cut potatoes. They learned to cut safely. In the meantime, they learned the tastes of all the things that came from our dirt, their smells and textures. There were some bandages as they learned but no stitches. Our pediatrician told us he rarely saw farm children for injuries. Despite the dangerous environment—or, rather, because of it, he said—farm children learn physical skills and boundaries earlier than most children, and they know the high stakes of transgression, because farm parents have to be sure to teach them.

Mark and I never shaded the dangers of our farm with gentle words. We never pretended there was no such thing as dead. "See that?" I told Miranda as I carried her around the farm on my back during the dangerous stage of childhood that comes between full mobility and development of a truly functional forebrain. "That's a bull. He's not gentle. If you go into his pen or his pasture, he will stomp you into the ground and kill you." Miranda was too young to keep up with me at a walk, but she knew well enough to stay far away from the bull, the boar, the ram, to move fast out of the way of horses on the farm road, to watch out for tractors, trucks, and machines.

The winter after we threw our boats in the lake, I turned forty-one. My friends Terry and Amelia came over for my birthday dinner. We'd been friends for ten years. Amelia's family owned a lake house near our farm, but she and Terry lived in New York City, worked in finance, and were one of my reference points for normalcy, tracking along the path somewhat parallel to the one I might have taken if my life hadn't made a U-turn at farming.

After dinner, during coffee and cake, Mark reached into the freezer and gave Jane a dead chickadee that had brained itself, earlier that day, on our kitchen window. Mark did not believe children should watch TV, play video games, or eat sugar. However, he saw nothing wrong with a child fondling a bird carcass during a dinner party. Jane held the bird at eye level, stroked its feathers, made puppet moves with the beak, the feet.

"Dad," Jane said, "could you cut the wing off, so I can see how it works?" Sure he could, and he did. Jane played with the dismembered wing for a while, then returned to the main carcass and slowly twisted off its head.

Me: "On a parenting scale of one to ten, one being perfectly normal and ten being we have to call the police, how weird was that?"

Amelia: "Two."

Terry: "Two."

They were such good friends.

The next day, Mark loaded us all into the car. "We're going to look at your present," he said to me. He drove south of town and pulled into an unfamiliar driveway, a tan house with a barn in the back. "This way," he said, heading for the barn. Over the crunch of snow under our feet, I heard the sonorous bleating of sheep. It reminded me of hearing full-throated Italian for the first time, in the airport, on my first trip to Rome. It was so rich with lyrical Italianness, with passion and emotion, I was sure they were putting it on. That's the way sheep sounded to me. So expressive of their sheepliness, so melodramatically plaintive and demanding. I threw my arms around Mark and kissed him.

For years, at our winter planning meetings, I'd brought up the idea of raising sheep, naming all the reasons I thought we should do it. They were small enough for me to handle and manage on my own. The children could work with me, and I wouldn't be so worried that they would be hurt or killed. They were ruminants that could be raised and finished on grass alone, without grain, which is always expensive, both environmentally and economically. They were delicious! The most flavorful domesticated meat in the world. Also, they produced wool, which could be sent out for processing into yarn, for another small stream of income. Mark would counter with equally strong reasons why sheep were a terrible idea. They were the world's most vulnerable prey animals. Coyotes would come from neighboring counties for the privilege of eating them. While they may be flavorful, it was a flavor the majority of Americas didn't tend to actually like. They were far more labor-intensive to raise and butcher, per pound, than beef, and in our one-price membership model, they would bring no premium. Sheep were delicate compared to cattle. Cattle have amazing immune systems, can recover from the most astounding injuries. There's a farmer saying: SSSS, for "Sick sheep seldom survive."

I got my way on my birthday, not because he agreed with me but because he loved me and wanted to make me happy. Also because of his thrift. A friend of Mark's had given us this flock. He had grown up in Greece and raised sheep as a hobby for thirty years but was tired of it and ready to get out. There were seven ewes, one with a lamb at her side and the rest on the cusp of lambing. The ewes ranged from yearlings to a toothless twelve-year-old, a sturdy American breed called Polypay, meant mostly for meat but with good, interesting fleeces that I have since heard compared to the fuzz of polyester batting, but in a good way. It spins into springy white yarn that knits up beautifully into thick warm hats or mittens.

We put them in the woodshed, which was empty of wood, and built them a small corral so they could venture into the frozen driveway. They lambed, one by one, singles and twins. The chorus of baaing greeted me every morning, as the kids and I walked out to feed and water them. They hadn't been sheared in two years, and the fleeces were extremely heavy. They were so well insulated, the snow piled up on top of them without melting. As the lambs grew, they used the ewes like giant puffy eiderdowns, standing on top of them, pouncing from back to back, then, when the ewes lay down, curling up next to them to sleep. The following year, I bred the ewe lambs along with the ewes, and the year after that, I bought in some new stock, saved the ewe lambs, and continued expanding. I learned as much as I could about sheep. In a few years, the project that was meant to be something small and easy that I could do with children would become a real flock of 250 head, and an important part of our business.

After my birthday, I packed the kids in the car and drove off to my hometown, three hours west, while Mark stayed home to keep the farm running. I spent the night with my parents, and the next day, I got up early and drove off, leaving the kids with my mother. They'd soak in some cartoons, sugar cereal, and grandmother love, while I went to visit our friends Donn and Maryrose.

They milked a flock of grass-fed sheep and produced a raw Roquefort-style blue cheese that made my knees weak with pleasure. They used draft animals for most of their work, like making hay, clipping pastures, logging and hauling wood in the winter. Donn had a fondness for draft mules, and he bred and kept some along with his Percheron and Suffolk horses and his mammoth jack, Eddie. I did not share his enthusiasm for donkeys and mules, but still, in the tiny corner of alternative agriculture where we resided, he and Maryrose were two of our closest neighbors.

They were also ten years older and more than ten years wiser.

Their farm was perched on a high grassy hill. I could see the sheep as I pulled into their long dirt driveway, which was bordered on either side with pasture. The light was returning—it was sugar season—but it was cold, and there were fat flakes of snow blowing horizontally through a pewter-colored sky. The ewes were heavily pregnant. I was glad they were wearing such good wool coats and were still a few weeks away from dropping their lambs.

Donn and Maryrose built their farmhouse themselves, with timber harvested from their own woods, and straw bales from the neighboring field. The exterior walls were plastered over with clay. I walked in, feeling, as I always did, the snugness of that structure, its security. It was modestly sized and, inside, both beautiful and entirely practical. The south side was dominated by a pair of large French doors that led onto an attached solarium, which was floored with heat-absorbing bricks, set in sand. There was a woodstove in the main room that doubled as a cookstove. On sunny days, even in the dead of winter, they would not need the stove. The greenhouse was the main source of heat, bringing a steady, radiant warmth into the house, which was so well insulated by the straw that it held the heat throughout the night. Whenever I got depressed about our house, I thought of theirs. Maybe I could build one, like they had, out of what we could grow and harvest on our farm.

I had arrived in time for breakfast. Every farm has its primary strategy for survival. Ours was a combination of diversity and intensity. Donn and Maryrose's was a strict sense of simplicity. They had imposed a quota on decisions. For example, so there was never the problem of deciding what to wear, Maryrose wore the same uniform every day: Carhartt overalls cut off just below the knee, over a T-shirt in the summer, long johns in spring and fall. In the dead of winter, she switched the Carhartts out for an insulated coverall. And instead of deciding what to have for breakfast every morning, they always ate exactly the same thing: black beans, some cheese from their sheep, and a fried egg on a tortilla. It was very satisfying. After the plates were empty, they picked them up and licked them clean to make washing up easier. I did not lick mine, so Donn reached across the table and licked it for me. "You can do it yourself tomorrow," he said, grinning.

When the dishes were put away, Maryrose reached for the bulk-size bottle of ibuprofen on the table and swallowed a big dose. Her knees, she sighed, were wrecked. Every Thanksgiving, as they went around their table saying what they were grateful for, she gave a shout-out to her good friend ibuprofen. She was also deep into what she called her menopause project, which, she said, was not for sissies. "The loss of muscle is the hardest to deal with," she said. She didn't have the strength she used to have. Jobs like shearing had to be done in steps instead of in one big day. This is one of the truths of farming. We are very ordinary athletes who can do what we do the way we do it only as long as our bodies will let us.

Maryrose and Donn had also simplified their roles on their farm, divided according to what suited them and their interests. Maryrose was in charge of sheep, grazing, and cheese. She had a degree in dairy science and was a master cheesemaker. She was nine when she went on a trip to Ireland with her family and saw someone hand-milking a cow, and she knew with extreme clarity what her life would be. And she had made it so. Donn was in charge of draft animals, haymaking, pasture, and people management.

I watched them move smoothly through their morning routine. Maryrose was walking a little stiffly between the table and the sink, waiting for her ibuprofen to kick in. "What will you do when you're too old to do this?" I asked. It was a rude question, but I needed to know. I had seen how work came to a full stop when someone was injured. But they were ten years older, and surely they would have the answer.

"We'll scale down," she said. "We can hire

people to help us. We'll strategize, probably move from wool sheep to hair sheep, so we won't have to shear." She paused. "And then, you know, we have a retirement plan." I leaned closer, because that was exactly what I wanted, and I had seen no way to do it. "We are going to run a donkey-drawn taco cart," she said, smiling at the vision. "Yeah," added Donn, looking out at the snow, which was blowing against the window. "Somewhere warm."

After breakfast, Donn and I walked outside to the barn. I looked over the backs of the horses and mules. Among his steady workers were always a few special projects. This time, it was Polly, a good-looking black Percheron mare who had washed out of a sleigh-ride business because of her tendency to bolt. Even in the corral, I could see she was high-strung for a draft horse, her energy coiled like springs in her pretty feet. Donn's interns, Scott and Daniel, walked up to her and hooked a lead line to her halter, then led her out to a clear spot under a group of trees. They would do some groundwork with her before using her to move logs from one end of the adjacent field to the woodpile. I settled my back against a tree and watched.

The object of the day's lesson was to reinforce, in Polly, the concept of yield. "Teaching left and right is easy," Donn said. "Yielding, acceptance, is the hard part." Yield relates to everything. We are small, slow, and weak, and they are big, fast, and strong. Try to lead a two-thousand-pound horse who has decided not to yield to you. Before Polly could learn to work happily, she needed to learn to yield. This would involve overcoming the instinct to push into pressure instead of moving away from it. Push your palm against an untrained haunch, and the haunch will push back and lean against you. This oppositional inclination, Donn said, is extremely strong in the donkey and weaker in the horse. The mule is smack-dab between them.

In order to teach yield, you have to apply pressure. Every part of a person holds the potential for pressure. Your eyes, voice, legs. Your very presence. It's a variable sort of power. Some animals are more sensitive to it than others, and some people carry more of it, naturally. The key to pressure, the thing that turns it into a tool for communication rather than a weapon, is the appropriate use of its opposite. Release. The removal of pressure is the reward for getting the right answer. Donn squared his eyes at Polly's flank, and she began to move around him in a circle. Her inside ear twitched front and back as she figured out what was wanted of her. There was no conflict in that conversation, only a gradual, mutual movement to an understanding.

Donn moved out of the circle, and Daniel

stepped in. He started at Polly's flank, but she didn't move. "Look at her eyes and ears to understand what she's thinking," Donn said. "Anticipate what she'll do. Set up ways to make the animal right. Avoid the power struggle. If she doesn't want to move, then ask her to stand."

Soon Daniel and Polly were in conversation too, not quite as smooth as Donn's but functional. Then Polly became irked with the small circles, stopped, and refused to move no matter what Daniel did. He sighed and looked to Donn for help. "You might as well stop there," Donn said. "If you are frustrated, you've lost."

We walked to the next field, where Scott and Daniel were supposed to hook Polly to a log and pull it out of the field. She trotted, and shied, and wouldn't stand still long enough for the boys to fix the chains. When they circled for another try, she rolled into a tight canter, and the boys struggled to keep up. "You have to make a distinction between her energy and your own," Donn intoned from the sidelines. "Don't get into a pulling contest! Don't be afraid of her energy. Just direct it. And figure out ways to make her right."

I watched Scott take the lines. He didn't know much yet, and this horse was a powder keg. But Donn was serene, arms crossed over his chest, spouting his koans from a distance. "Your energy is not her energy," he boomed.

There were moments when I wanted to grab the lines, sure the boys were going to get crushed by the log or lose control of Polly, but Donn busied himself with clearing away branches, chatting with me about mules, one eye on Polly, shouting corrections from afar. Making a distinction, in other words, between his energy and theirs. At last, Daniel unhooked the tug chains from their place at the hip of the harness. Scott asked Polly to back until the tugs reached the log. Daniel clipped the chain on, and the horse and the tree were joined. She danced, testing, testing, looking for a reason to escape, and when Scott told her to come up, she hit the end of the chain—BAM!—and the log moved. Polly dug in, tucked her pretty head, all business, and pulled.

At lunchtime, I found Maryrose on her back in front of the washing machine, tools scattered around the floor, which was freshly mopped from a flooding. She pulled off the cover, fished around, brought out a plug of hay, straw, and horse hair. I'd done the same thing with my own machine the week before, extracting a wad of similar composition. Machines in the civilian world are not built to withstand what farmers throw at them. Maryrose had spent her morning on sheep chores, turning the cheeses in her cheese cave and cutting labels for the rounds that were ready for market. Unlike Donn, she preferred to work alone. This was understood between them.

I spent the night with them, then drove to my parents' house and picked up the children to head home to the farm. The girls conked out in the back. In the quiet car, I thought about what Donn had been teaching and how it applied to marriage and parenting. The way that in a partnership, the idea of yield is everything. In order to work together as a team, you had to first overcome the natural urge to *oppose*. Instead of addressing all the ways it's gone wrong, look for ways to make the other one right. And when things get wild, dark, or hard, make sure to separate your energy from the other person's energy. *Your energy is your energy,* I intoned to myself, *and his is his.*

CHAPTER 14

I kept a stack of books beside my bed. Straw-bale house design. How to timber-frame. I read up on skills and tools. I invented a house in my head that I could build, away from this one. Why couldn't I make a house for us from the farm itself? Bright and snug? We had wood for the timber frame, stone for the foundation. The walls could be made of straw and clay. I read all about it. Rye straw is the best for walls. When we planted it that spring, I watched the rows germinate. As it grew, the girls played house in it, and I imagined it as our actual house. When it was cut and dried and baled, the straw was stacked in the East Barn run-in, bright yellow, clean-smelling. I'd look at it on my way to check on the chicks or collect eggs and think about what it could be. But then we needed it, urgently, for bedding for the dairy cows. It disappeared bale by bale, shat upon, swept into the compost pile, until it was gone. Meanwhile, the children were getting older. You can see time passing very clearly as your kids are growing up. I clocked how much of their childhood was left, how much had been spent in the dingy old house.

When Miranda turned three, I felt an old familiar restlessness that usually presaged a change. In another time, it would have been answered by heading off to a new country or a new relationship. The core of me is a little warped in that way, its tentacles always stretched out, seeking novelty. I had to learn how to manage that as a married person with children and a farm. I knew I didn't want to change the foundation of what we'd built together. Especially now, when it was solid. But maybe we could bump out a wall, build a little addition? I asked myself, as candidly as I could, what was it I wanted? It was not another child. We were too busy and quite possibly too old. But it was something new and small.

That spring, we'd had a visit from Houlie, who

used Jet as a stud dog on her English shepherd, Rosie. Jet got a lot of work as a stud back then. He was larger than most English shepherds and had a rock-solid temperament that made him easy to work with. Rosie was a star where she lived in Pennsylvania, a small, lightly built search-and-rescue dog trained in air scenting, trailing, and cadaver work. She could work stock too; Houlie and her husband kept goats, chickens, turkeys, and sheep on their farm. But Rosie was so keen to work, and so intolerant of foolishness, that she needed a skilled handler. Houlie was a professional dog trainer and described Rosie as a canine Ferrari—an awesome animal to own and work but too much dog for most people. Houlie thought Jet might add some size to the line and moderate Rosie's ambition enough to make her pups more user-friendly. I trusted Houlie's judgment. Stud fee can be taken two ways: one pup, your pick of the litter; or a pup's cash equivalent. Back when Rosie was bred, I was certain we wanted the money. Now that the pups were on the ground, I wasn't so sure. I called Houlie, who was sorting through puppy contracts, registrations, buyer applications. I wanted her to talk me out of it. "I'm not sure I can manage a pup right now. Another infant."

"Yes," she said, "but each stage only lasts days or weeks, not years." I took that in and considered.

• • •

Mary arrived on a fine fall day, when the girls were three and six. Houlie had picked her for me after getting a look at what our farm was like. She knew the stresses, the daily challenges, and also the opportunities. Working breeds like English shepherds are rare because the types of work they were bred to do have become rare. A diversified, mixed farm like ours was where an independent-thinking dog who could hunt, herd, and guard could really show her stuff.

We needed a dog, Houlie said, who was brave and could withstand frequent challenges. That sounded to me like a good set of credentials for anyone who wanted to go into farming. Houlie watched as each pup from the litter encountered the electric fence on her farm for the first time, in the course of exploring the world. The pop of a fence sends some pups into blind panic, to huddle at the handler's feet, or all the way back home, to the place they feel safest. It wasn't the reaction that was important but the recovery. Houlie watched Mary hit the fence, yelp once, and come right back sniffing in circles, looking for what had bitten her, a little more cautious but unafraid.

It was clear from the beginning that Mary had gotten a full dose of her mother's genes and not very much of Jet's. Even as a little pup, she had raw ambition. Arrogance, even. And was so tolerant of pain that she could get kicked hard by a frisky heifer and jump right back in the game. Her motivations were more complex than Jet's; she had her own agenda. Unlike Jet, who was born honest, Mary was capable of duplicity. Once I made a stack of pancakes for breakfast and left them on the table. I went upstairs to wake the girls and came back to find we were now one short. Mary trotted out of the playroom and lay down by the stove, where I had left her, sleeping. *Hum-de-dum-dum,* her eyes said. I walked into the playroom, rooted through the children's toy box. There, under a layer of balls, dolls, and plastic horses, was the pancake. Mary watched from the other room like an actress cast to play innocent.

Jet worked out of obligation and a strong sense of duty. Mary worked for the absolute joy of it, with her whole heart, the perfect clicking of her instinct into action. About the only things I recognized of her father in her were Jet's sharp intelligence and his capacity for loyalty. Mary tolerated Mark and liked the children fine but lived with one eye on me, in case I moved toward my jacket and boots. She was always hoping we had a job to do, preferably a big one that required her to run fast and use her teeth. There weren't many jobs like that, in reality. The key to using a dog on stock is to keep things calm, low stress for everyone. Where Jet had needed encouragement,

Mary needed to develop restraint. There was no way I could have handled her if she had been my first working dog; as my second dog, I could, but barely. She was a lot like my husband: challenging, enthusiastic, and never dull.

I cooked, took care of the kids, trained the puppy, who went everywhere with me, tethered to my belt by a light rope. The crops came on, and the weeds. Farmers left, new farmers arrived, a whirl of youthful energy. Many of them are smudgy images in my memory, faces who came and went; others stayed for years, became an extension of our family. They learned skills that would serve them later on in life, whether they kept farming or not. Hard skills, with their hands, and soft ones, of character.

The strings between me and the girls loosened, a little at a time. Jane began exploring the sugarbush, the swamp, the edges of the field on her own, or dragging Miranda along behind her. They went a little farther afield every day. Sometimes they hiked out to pick strawberries together, coming back with their faces covered in juice, their baskets empty. From May until September, they were almost always barefoot. One afternoon they came running into the house, shouting with excitement. Jane had found a snapping turtle hatchling, the size of a silver dollar and the color of the muck at the bottom of our pond. It stretched out its tiny wrinkled neck, showed its beak. That beak, when fully grown, could take off my thumb. Jane put the turtle down on the kitchen table. They claimed her for a pet and agreed to name her Charcoal. They watched turtle videos on YouTube and learned that she could live for a century and grow to forty pounds, and if they wanted to keep her they'd need a heat lamp and an enormous tub of water. So they decided she was only a twenty-four-hour guest, after which they'd return her to the puddle she came from.

This was first in a line of not-furry creatures that the girls befriended. Snails, spiders, and a terrifying venomous water bug had all done time in my kitchen. Jane was a master at catching snakes but not so great at holding on to them. Two of them had escaped inside the house and slithered into the dark unknown, never to be seen again. I hoped they found each other, in one of our crevices, and escaped the house together to start a little snake family of their own.

In the fall, Jane went to public school. She was fine getting on the bus, she had a ball in class, and she loved meeting new friends. She had one concern: lunch. Why did she have to pack our usual homegrown stuff in her lunch box while her classmates brought bright packaged "things from the store" that she hadn't learned the names of yet? *Because food is the center of our family,* I thought. *It governs the rhythm of our day and of our year. It is our sustenance, our living, our lifestyle. And because that pork belly and sauerkraut sandwich I made for you totally* rocked. Then I tried to translate that into language she would understand. "Food," I said, "is what we do. It is one of the things that makes our family special." She took that in, thinking.

Miranda went to preschool, and a space opened up again in my days for the farm. The New York City share was growing fast now, bringing in more revenue than our local share. We'd hired more staff. The whole business was big enough that I had the luxury of being able to choose what I wanted to work on. I chose the sheep and the dairy.

When the grass began to grow, I went to see Racey's new farm. She and Nathan had bought a beautiful piece of land in the valley to our west. Chad and Gwen bought another piece right across the street. The fields looked like ours had when we first started. Disused, rough, full of potential. They were ringed with hills and already had horses on them: Nathan and Racey had bought a trio of Suffolk weanlings, two fillies and a colt. Their coats were the color of new pennies. Chad had his eye on a Suffolk stallion. They'd work together, breeding and training horses and raising vegetables, grains, sunflowers, and meats.

Racey and I walked through the fields, talking about soil, which animals or crops might thrive in each field. The sadness and anxiety I'd felt when she left our farm was gone entirely. Mark was right. More farms were a good thing. We could cooperate instead of competing. We all needed to stay nimble, anyway, keep looking for new markets. And it was better than good—a great thing—to have my friend on her own land, nearby.

The land came with an old house, which was slipping off its foundation, already condemned. We walked inside. Nathan was at a makeshift desk making a list for the day. The room was big and bare, with cracked windows and a plywood floor. They had set up a camplike kitchen at one end, a giant table in the middle, the office at the other end. They slept upstairs in a loft accessed by a ladder. The house had been damaged in a fire in the 1970s and never properly repaired. The exposed beams were charred, and some daylight showed through the walls. Racey said that during the winter, when the woodstove died overnight, the temperature in their bedroom had dipped regularly into the twenties, and the glass of water she left by the side of the bed froze solid after midnight. They'd be building a new house, but it would take a while because they had a barn to build first, and they'd be doing all the work themselves, in their hours away from farming.

This hard living in the condemned house was part of the plan. They were lifting a new farm off the ground, and this was what it would take. They were working their way toward ownership, as we had done ten years before. There was some urgency, though, because Racey was going to have a baby.

There is a poem by John Berger about an old peasant visiting with Death, who will soon claim her.

> *What she asked him was his opposite?*
> *Milk he answered with appetite*

At the end of November, I walked through the barnyard in the dusk. One heifer had just freshened, and I was expecting another one, Posy, to calve any minute. When I had checked her at lunchtime, everything about her looked big. Her body was full of calf, and her bag was very tight and swollen.

It wasn't officially winter yet, but the weather had already turned cruel. There was a dusting of hard dry snow on the ground, and the wind picked it up and threw it at my face. The forecast said it would be zero that day, below zero at night. The cows were in the barn already for milking, and I slipped in at the heavy sliding door, with Mary tethered to my belt, and closed it behind us. The wind screamed around its edges, but inside, it was warm and quiet, the only sounds the music of the cows' bells and the rustle of their eating.

The dairy had been in a long period of quiet anticipation. Most of the cows were dry. This is the rhythm of a cow's life: She calves, which turns on the faucet of her milk. It will run, with varying intensity, for the better part of a year, until it slows or she is two months away from calving again and we stop milking her, thereby turning it off. When she is dried off, we keep her in the milking herd for a few days, to rub her udder with liniment and make sure she doesn't come down with mastitis. That seldom happens on our farm; by the time we dry our cows off, they are producing very little milk. After the cow is dry, she is turned out to pasture for two months to rest and regain condition before her next calf comes, which turns the faucet on again.

The dry period had been longer than usual that year, and we were calving in the cold dark edge of winter, because of a bull problem. The bull had come on loan from a dairy in Massachusetts. He was supposed to be proven as a sire, but when we got specific with the farm who rented him to us, we learned that he was used to breed cows in a herd that included other bulls, so he was not proven after all. He seemed like a good enough specimen. I'd watched him trot off the trailer and into the pasture and immediately begin the work of a bull. But after several weeks of diligent breeding, I noticed the cows who should have been settled were coming back into heat. They weren't pregnant. If it had been one cow or even two, the issue would have been muddled; I might have suspected a cow problem. But if all of them were coming back into heat, the problem was the bull. The trouble was, it was hard to be sure. I put Miranda on my back and went out to investigate. We bicycled to the barnyard, where the bull was with the cows.

There are no biological secrets for farm children. When the bull jumped on the cow's back, we maneuvered in close, one eye on the thrusting bull, one eye on a means of escape should he turn on us. "They're *breeding,*" Miranda said, sighing. "That's good, right, Mom?" Everything looked routine at first. The cow was in strong heat. She stood eagerly, the bull mounted and thrust, and I could see the bright pink stick of his penis. But upon looking closer, I saw that it wasn't like a normal bull's, not straight but curved sharply to the right. When he thrust, it was not hitting home but brushing bizarrely against the cow's flank. I biked home with Miranda and Googled "bull penis deformities," which nobody should do on a full stomach. This guy had a lateral penile deviation. It might have been congenital or caused by injury, but in either case, none of the cows had been bred. It took a few weeks to acquire another bull. So Posy was calving in the cold of late November instead of the good fall grass and gentle sun of September, as we had intended.

I walked through the barn with Mary. The wind had built to a gale, with sharp snow in it. It was a little more comfortable in the covered barnyard, which we had built the year before with a grant from the USDA. It was airy, deeply bedded, and had lights. We had added tarps as walls to break the wind. The USDA will give you a roof and a cement pad to keep the runoff out of the water, but it won't give you walls, because that would be a production advantage, keeping your cows warm. The tarps were free, and they worked.

There were five dry cows and heifers. Four of them stood around the bale and ate. But Posy was by herself, lying down, standing up, pacing, stopping for a mouthful of hay every once in a while. She was in early labor. Her udder had changed in just two hours. It was so tight now it looked almost translucent, her little heifer teats standing straight out at the corners, a bit of blood on the ends from burst capillaries. She was a beautiful cow, shaped like a fat triangle, and so ripe with her calf. Mary sniffed in the bedding, alert. Then Posy swung her head toward us, saw the puppy, and tensed up. She wasn't in the deep part of labor yet, when the rest of the world disappears. And this was her first calf. I didn't want to worry her, so I took Mary home.

When I got back to the barn, Posy was deeper into her labor. She lay down and then stood up. I sat quietly in a corner to watch. The small feet—one hoof brown, one hoof white—poked out of her, retreated, poked out again. The contractions came over Posy, closer and closer, stronger and stronger. The other cows wandered by and snuffed at her, then returned to the hay. Posy did not want their company. She smelled the straw where the birth fluids had dripped. She wanted that spot to herself. Finally, she lay down and did not get up again. The other cows left her alone. She was completely inside her own self. Her neck stretched out with effort. Her mouth opened, and her pointed tongue curled up to touch the roof of her mouth. She let out a deep moan. The little feet advanced again and did not retreat. I had a terrible impulse to interfere, to grasp those feet and pull. I knew better—as long as there aren't complications, it's safer for the cow and calf, to let birth happen at its own pace—but that was my instinct.

The colors of birth are blue, white, and red. It is a wet event. The calf is a strange and perfect package, tight and efficient, wrapped up in its sac, pointed toward the exit. Every once in a while, the calf would move a foot. I wondered what Posy was thinking. First calves come slowly, the Old Testament business of the firstborn, who wrenches open the womb. Did Posy think she was dying? Did she think her body was expelling some large, important organ? Maybe there were no thoughts. Steadily, the package advanced. Now the head was out, tucked between the knees. Another tremendous push, and the shoulders were born. Then a slight rest, panting, and one more massive effort, and the calf slid out, whole, separate, and free. The cow lay unmoving, openmouthed, panting, panting. The calf moved first. It wriggled and unfolded, seeming to bloom from a wet arrow into a calf. It bleated. One ear flopped free of its head, where it had been pinned by the funnel of birth. The cow's consciousness returned, and with it the recognition that there was another being on the straw behind her. Some ancient piece of understanding clicked into place. She mooed the low sonorous moo that is heard only at births. She stood, and the cord between them snapped. Then she turned and, with surety and purpose, began to lick her baby. She licked with vigor, as though the calf were the most delicious thing she'd ever tasted, as if she wanted to imbibe it.

After she had finished her cleaning, I walked to them, to check. If it were a bull calf, the joy would be dimmed by knowing he wouldn't be with us long; if a heifer, the joy would be full. A heifer is milk, the opposite of death, and she would get the best of everything—the most attentive care, the best grass and feed. And if all went right—if she grew and thrived and got pregnant and had a calf of her own—she would live a long and productive life. I walked up to the wet creature, one ear still pinned, body slick and steaming in the cold air, and felt between her hind legs, and found the four nubs of her future. A heifer. It was cold, and the wind was blowing so hard that the drafts in the barn were like confined storms. The heifer shivered a little at the cruelty of being squeezed out of her floating waterbed to this harsh, frozen world, eighty degrees colder. I let Posy work on her for a while longer. She and her tongue would do better than I and my towels. Then I went to find blankets and to add straw bedding to the calf pen.

It's always hard to separate them. I'd felt the cow's labor in my own body, a faint echo, the mammalian sisterhood. When she pushed, I felt the floor of my pelvis release. When she opened her mouth, the tip of my tongue found the roof of my own mouth. As the shoulders emerged and the vulva stretched, I felt impossibility, desperation, inevitability, then resignation to the fact that the only way out is through. And so I had to close my eyes when I took the calf away because I'd feel that too—the longing, the not-rightness of it. I couldn't avoid it. It had to be done. I picked the calf up and slung her legs over my shoulders, the wet noodle of her umbilical cord cold against my cheek. I lured Posy into the barn, using the calf as a magnet. She followed with that low, grunting maternal moo. I put the calf down in front of a stanchion, and Posy put her head through to sniff her. I locked her in, ready for her first milking, then picked up the calf and carried her away to the blanket in the calf pen, where another heifer calf was napping, only three days old.

Back at the house, I reported the birth to Mark and the girls. Mary examined the smells of birth on my clothes, intensely curious. It was hard to tell what the emotion was behind the sniffing, but it was eager. Mark went out to milk Posy, and I gave the girls their dinner, then we switched places: Mark got them ready for bed, and I went to take care of the calf. Mark had left a gallon of colostrum in the milk house and covered the sleepy calf with a blanket, half buried her in a thick layer of clean straw.

Mary was with me again, not tethered this time, just trotting free at my heels. She was young enough to be scared of the dark and not stray too far from the den of our house at night. In the nightly audio war against the coyotes, Jet was on his own. Mary barked her tiny bark from the doorway. But the combination of the spot of light from my headlamp and that intriguing smell on my pants gave her courage, and she came along. She was growing up.

Two bottles of colostrum steamed in a bucket of warm water in the milk house. They were thick, and the usual yellow-orange was tinted bright red, full of blood from the burst capillaries in Posy's swollen udder. There were drips of it around the edges of the bottle. I carried it inside my coat to keep it warm, and the colostrum dripped stickily down my hand. Mary, who knew she was not allowed to jump on people, jumped straight in the air next to me, trying to get closer to that smell.

When we got to the back of the barn, the calf was standing and had shed the pile of straw and her blanket. She looked alert now. I climbed over the rail of her pen, and Mary flattened herself to go under. The calf was too young to be frightened of dogs or anything else. Mary sniffed warily at the calf, then rooted through the straw and found a frozen black puck of meconium, her first feces, the remnant of everything that had happened in the womb. The word "meconium" comes from Greek and means poppy juice. When a calf is born unobserved, you can tell it has sucked by the color of its feces. A calf with a belly full of colostrum will drop bright orange blobs. These too were scattered around the pen, compliments of the other calf. They were half buried in straw, and when Mary discovered them, she found them even more interesting than the black puck.

The power of colostrum is temporary but vitally

important. It would give the calf immunity to the pathogens that her mother had encountered in her lifetime. There is only a short period after birth when the calf can absorb that magic. The first hour after birth is best. Six hours is okay. In twelve hours, the open spaces in the calf's digestive system that could absorb those big protective molecules would have begun to close. It made me think of that liminal time of childhood, the one Jane was exiting, that Miranda was firmly in, when the wall between real and imaginary is permeable. Through the wall fly fairies and spirits, until time seals it up and they become fleeting memories.

I straddled the calf with her head just in front of my thighs, the bottle in one hand, her chin in the other. I wet my fingers with the colostrum and pushed them into her mouth. She sucked them, stepping back as if surprised that something so good existed in a world this cold. I walked backward with her, my fingers in her mouth, until she had backed into the corner of the cinder-block walls. Then I lifted the bottle and pushed the big rubber nipple into her mouth, still holding her between my legs. I pressed down gently against her forehead so she would drop her neck and stick up her nose, to open her esophageal groove and send the colostrum to the proper stomach. Some calves take a few days to learn to suck, but this one was sharp from the start. She sucked, and sucked, and sucked. In a minute, she had downed a quart of colostrum, and the sides of the bottle were caved in with the pressure of her sucking. I pulled the nipple out of her mouth to let air into the bottle, and she resumed her feeding. It is astounding how much colostrum a newborn can drink. I imagined it filling and unfolding her stomach for the first time, like the air had filled and unfolded her wet lungs after birth. She sucked vigorously, a quart, then two, still going. She wouldn't drink this much milk again, even when she was bigger.

After the calf started the third quart, still sucking hard, I heard something faint and then disconcerting. I looked down to find Mary chewing and tugging on the umbilical cord, which was fresh, fleshy, and attached to the calf. I shooed the puppy away with a growl and a stamp of my foot. Boundaries! She'd need to learn. There is prey, and there is pack. Calves were pack, part of us. In order to underline the point, I ignored Mary and fussed over the calf, stroking her forehead and cooing.

When the second bottle was empty, the calf was so full of warm colostrum that she trembled. Her tongue pushed the nipple from her mouth, and she collapsed into the straw. There was almost a gallon in that little belly. That was good. Without colostrum, there's not much hope for a calf. With a little colostrum, it would be an uphill battle against illness—scours, navel ill, pneumonia. With this much in her belly, she could face all sorts of pathogens, and I would not worry about her in the night.

CHAPTER 15

Winter arrived, and we settled in for the long haul through the dark. The calves were growing, my flock was growing, the puppy was growing, the children were growing. The farm was growing. Slowly, we came back from the precipice of insolvency. We paid our bills, the mortgage, the taxes. We owned some of the land free and clear. We had assets but no comfort yet. The house was still dim and dark. As winter began to shift to spring, Elizabeth came to stay with us. She was dating a farmer named Paul who had worked for us and then started his own farm a few hours away. Elizabeth had a degree in philosophy and a strong socialist bent. She'd gone to work at Paul's place as a summer farmhand after an unhappy stint at a chocolate factory. Elizabeth was smart, focused, and intense, and she was having a hard winter. She and Paul had butchered thirty-two pigs between frost and Christmas. There was only one other person on the farm during that time, and three people shouldering all the usual work plus the butchering was simply too much. They'd worked insane hours. When the last pig was cut and in the freezer, Elizabeth went down. She stayed in bed for ten days and spent them, she said, alternately reading and crying in a wet, messy, uncontrollable way. "You must have needed a rest," I said. "It was not a rest," she answered. "It was a breakdown." It was brought on by too much work and too much stress, by trying to figure out all the details of partnership, farming, finances, the shape of their future lives, all of it rolled up into one big blunt thing that knocked her over. She was at our farm to get some perspective. She wanted to know what the view was like from where I stood, another decade on.

I'd invited Racey over to join us for tea. An hour before they were due to arrive, I pulled out a bag of flour, mostly white, that I'd sifted from the bag of hard red winter wheat that Barbara had ground fresh for the share the week before.

The wheat had grown in Superjoy, in that good drained soil. I remembered watching it sprout in the late fall and come back, after it was grazed once, to grow to its full height, then turn brown, the heavy heads bending under their own weight. Once wealth was measured by the whiteness of food. Brown things, *unrefined* things, were for poor people, white things for the rich. In our house, white flour was for company, not because it was expensive but because it was refined, too easy to digest, and too poor in nutrients for everyday eating.

I put a pint of butter on the back of the stove to soften just a little. This butter and I were old friends too. I had milked the cows and separated the cream, churned the cream into butter, and packed it into pint jars. The preciousness of butter, of all fat, is lost on the world. This pound of butter was made from twenty pounds of milk, half a day's production for one of our mid-lactation cows. When the butter was soft, I mixed it with sugar and a splash of vanilla from a nearly empty pint jar stuffed with fragrant whole beans and refreshed every few months with a slug of rum or bourbon. The beans came from our friend Rick, who had bought a train car full of them when the price of vanilla beans dropped. He parked it, waiting for the price to rise again. He brought us a whole brown bag of them and handed it to me in our kitchen. I opened it, and the room filled with their dark, blooming, spice-island smell. I suddenly felt like my kitchen was richly adorned, as if he'd handed me a bag of rubies.

I added the flour to the sugar and butter and mixed it into a thick dough that I pressed into a pie pan and pricked all over with a fork. For fun, I scattered a few pieces of lavender on the top, from a jar that the girls and I had gathered and dried the previous summer. Then into the oven it went.

Racey called hello through the kitchen door. Her belly was so big, it preceded her down the hall. Her baby was due in the spring. We poured tea and ate shortbread, and while the children scribbled in the corner, we talked. Three women, from three different farms, at three different stages of life, talking shop. I realized, a little startled, that *I* was the elder. When did *that* happen? Elizabeth was looking to me, to give her a preview of what was to come, in the same way that I had looked to Maryrose. Racey was looking to me too, to get a glimpse of motherhood. I wasn't sure I was qualified, but who else was there? Elizabeth talked about the overwork and the way the farm always came first. She worked hard, but compared to her, Paul was a maniac. When he was stressed by a problem on the farm, he threw work at it. And worked and worked and worked. He would work beyond any reasonable person's idea of how much one should work. This scared her. Given that she had hit her limit, she wanted to know, should she continue to work on the farm? Should she get an off-farm job? "Should we," she asked, "get married?"

I was glad Racey was there too. She and Nathan had had their difficulties. Show me a couple who hasn't. They had just finished building their barn and were rushing to finish their house before the baby came, before the spring work hit. They had already moved out of the old one into the new house's basement. It was a balmy 55 degrees in there, she said, and she couldn't quite believe the luxury. She had lived two winters in a house so cold they couldn't take off all their clothes at once. Two years without being able to take a shower whenever she liked, instead arranging a time with a neighbor. There were some glitches: the construction had created a mud pit through which she must wade, heavily pregnant, in order to leave the house or return to it. Also, the shower that she had awaited so long used an imperfectly repaired secondhand heater that shifted to cold without warning, so someone had to be stationed next to it and flip it back on mid-rinse. But the size of the improvement was immense. I asked Racey what she thought the best part of farm life was. "We become extremely easy to please," she answered.

Most of the questions we talked over that day

weren't about agriculture or our challenging living spaces. They were almost all about the relationships. How do you share power, decision-making, and responsibility on your farm? How do you divide the work? How do you shape the thing you love without fighting over it? How do you find time for love? How do you have anything for yourself? How do you stop the farm from eating you up? How do you work in a professional way alongside the person with whom you share a bed and crazy amounts of emotional subtext? If you have employees who live and work closely with you, where is the line between boss and friend? And how to maintain peace with those people while also being an effective manager? How do you make boundaries and construct some privacy for yourself?

I felt for them but found their questions comforting. They reminded me that these problems weren't just ours. This was human nature, and this was what it took to lift a farm off the ground. We were responsible for our own happiness and could weigh the costs against the benefits, and then we would have to choose.

Mark came in from the barn and sat down with us. For someone who could be astoundingly obtuse about our own emotional conflicts, he was often insightful about others'. He peppered Elizabeth with advice: If it's not working, redefine your role. Trim your hours.

"Rest enough," I added. Drink water. Exercise. Meditate. Take baby steps. Make a list of doable things. "Remember that you control what you choose to believe in," I said. Anyone could look at our farm lives and see toil and squalor, or look at the same lives and see purpose, abundance, and joy. Same plot points, different story. "Very different things can be true simultaneously, and choosing the one with the better narrative is often extremely helpful."

Elizabeth asked what had made me decide to commit to this life. I said that farming is not easy, and it is not going to make you rich, but it is a strong organizing principle, the best one I had found. It makes everything else fall into line. Everyone needs an organizing principle. Some people have money, or status, or art, or fashion, or church. We have farming and food. The axis around which our family revolves is a heavy rod of daily need: the plants need, the animals need, the farm needs. That's not a curse but a blessing.

At the end of winter, when Mary was eight months old, she went missing. I didn't notice at first. Donn was visiting, and we'd spent the day training horses and new teamsters. After dinner, I walked him out to say goodbye and nearly tripped over Jet. He was standing broadside to me in the mudroom, as though he'd been waiting for me to come out. His head was low, his tail down, and his eyes were locked on mine. Jet was far too dignified to make himself a nuisance over nothing. What in the world was he trying to say? That was when I noticed that Mary wasn't there. That was weird. She, like Jet, was a stick-around dog, and even at a young age, she had predictable habits. At that time in the evening, she should be there, on her bed, in the mudroom.

Maybe she was searching for butcher scraps in the compost pile, or hunting possum, or chasing Penelope, the infernal cat, or getting into some kind of trouble. That seemed a likely possibility, actually. Jet liked rules and wanted every beast on the farm to obey them. He could be a tattletale. I walked outside and called her. Mary was enthusiastic about many things, but nothing more than coming when I called her. She would leave what she was doing and arrive with great joy, at full speed, to sit in front of me, wagging her entire body. But she didn't appear. I listened for her bark. There was only the wind.

It was getting dark. The weather was turning bad, a sleety snow blowing out of the north, threatening to turn into a real tempest. Earlier, Mary had been with me in the machine shop. It occurred to me that if the wind had blown just right, it might have blown the door open, then closed, and she could have gone looking for me and gotten stuck. I ran over there, opened the door, and called.

Meanwhile, Jet was by my side, eyeballing me. "All right," I said. "Where's Mary? Go find her." He ducked his head and started off at a trot to the north, glancing back at me every few yards to make sure I was following. I followed him past the granary and the East Barn, pausing to open every door and see if Mary was stuck inside. I checked the compost pile and the hole next to the East Barn where the dogs stalked rats. Every time I stopped, Jet waited patiently and then struck off again in a curious side-to-side gait, always to the north.

When he rounded the west side of the barn, he broke into a run along the farm road, past the wrapped hay bales, past the turnoff that leads uphill to the sugarbush. That, I hadn't expected. I decided that he must be either blindly searching or thinking we were going for an evening walk together. There was nothing beyond the turnoff except pastures and, a quarter-mile away, Matt's old cabin, which hadn't been used since last summer, and in another half-mile, the covered barnyard where the horses and the beef cattle were spending the winter. I couldn't think of anything that way that would attract Mary or trap her. Meanwhile, it had grown dark, Mark was working in the butcher shop; the kids were supposed to be in bed; and calves had to be fed. I turned around and went back to the house. Jet followed far behind me, reluctant.

I found Mark, to put him on kid duty while I searched. "Have you checked the back of the machine shop?" he asked. There was a second room in the back of that building, where we kept the woodworking tools, and he was right, I hadn't checked there. It seemed likely. I ran back. I was really worried now. I burst through the second door, into the woodshop, expecting to find her, but she wasn't there. When I came back outside, Jet was sideways again, directly in my path. His expression had shifted slightly. He was always a gentleman, a diplomat, but I detected some impatience in his eyes. He was saying: *This way, dumbass*. But respectfully. He had very nuanced eyes. "Okay, fine," I said. "I'll follow you."

He struck out north again, faster this time. I ran, trying to keep up. Around the West Barn, north along the farm road, past the wrapped bales, past the sugarbush, past the road that led to the neighbors', down the hill, toward the faraway pastures. When he got to the cabin, he bounded through the deep snow and stood, front feet on the steps, and looked at me. His expression was almost pity at this point. *Poor human, is this sufficient, or do I need to actually spell it out for you?* I ran toward the cabin, noticing finally that there were fresh boot prints in the snow, leading to the door. In the last bit of evening light, I saw movement inside. It was Mary, leaping straight up, paws bonking against the cabin's window.

Mark had taken some visitors to see the cabin that afternoon and hadn't noticed that Mary had slipped inside. Apparently, Jet had.

Mary was coming along as a herding dog. I'd never learned the sort of impressive skills that you see in sheepherding competitions. I just brought Mary with me, and together we figured out how to get things done. We practiced on the dairy cows, walking them in for milking, and then on the sheep, keeping it simple and safe. Where Jet was large-boned empathy and soft power, Mary was police tactics, force. She was small, fast, judgmental, and more useful to me in the field at a year old than Jet ever was. But some things made for complications. Her instincts to guard and to please me were very strong. She was always watching, and sometimes I taught her things I didn't intend. She picked up on my need for boundaries, for example, and decided to defend them for me. When one of the farmers knocked on the door while I was in the house, it was usually because there was a problem on the farm—animals out or injured, or the well gone dry. "Uh-oh," I'd say under my breath, wondering what type of complication was coming my way. That was the same thing I'd say when something went wrong while we were bringing in the cows and a heifer stepped out of line to sneak a mouthful of chicken feed, or stopped to graze a sweet patch of clover. It was Mary's cue to streak in and enforce my rules with a quick nip at the cow's heel. It took me a long time to figure out that this was why Mary was nipping at the heels of the farmers when they came in the house. It took even longer to train her out of it.

Then it was spring. Lambing time, three years since the first ewes arrived, and the flock had grown from seven ewes to fifty. Daytime highs hit seventy, and there was a fresh wind blowing the water off the fields. The first greens appeared among the browns. The birds were in overdrive, building. My nightly checks in the lambing barn left me stumbling back to the house with a crust of colostrum and afterbirth on my hands. I fell asleep on the couch, and the dog licked them clean, then I woke up, ready to go. The warm days brought a physical sensation that felt akin to looking at a freshly pulled double espresso waiting on the bar. It was the same anticipatory rush. The body feels it before it hits.

The first lambs to come were twins, a ewe and a ram, to a ewe I'd bought the year before, a first-timer. The lambs stood right away but struggled to find the teat on wobbly legs. The ewe swung away from them, confused. I pinned her against the wall and held the little ram lamb's mouth to her teat. When his sucking mouth found it, I felt his body relax with the rightness of it. Then his tail began to flit back and forth, a metronome of satisfaction. By the time the ewe lamb was fed, the ewe had the hang of it. She nudged her lambs and nickered to them. I put the three of them in a small pen together to bond, rest, nurse. After two days, when they had gotten to know one another, I'd release them into the larger flock. But first I weighed them: 7.5 pounds and 8.5 pounds. Just about the size of our own children at birth, Mark pointed out. Could these two strapping children have ever been so small? I know they were, just as the ewe, who now outweighed me, had been this fragile only eighteen months ago.

The next day I watched one of the ewes give birth. She'd been so big that we all guessed it would be triplets, and out they came, 7.5, 9, and 10 pounds, a real belly full of lambs. It was a busy hour for that ewe. She was an old-timer and knew her business. Even as the second lamb was emerging, she was licking the first, and as the third came, she was singing to the first two, reaching for them with her tongue. She nudged them each to her teat and let them suck.

Racey had her baby, a beautiful little boy named Lewis. I sat on the edge of her bed in the basement of the new house and held him. He was four days old, a tiny mite of a human. Racey looked overwhelmed and very, very tired. Her birth had been hard, the price of the firstborn. Her eyes filled with tears as she said, "Of course nobody could have explained to me how intense it would be," and there was the barest hint of betrayal in her voice, an unspoken question—why hadn't I warned her? I gathered their laundry and brought it home to wash, like Ronnie had done for us after Miranda was born. Nathan had snapped to, taking on the role of father, rushing in from chores to hold his son, make tea for his wife. But she was right. Nobody could have explained how it would be.

CHAPTER 16

When summer came, we had new farmers in the field, new horses. We were still doing most of our fieldwork with horses, but we were often short on trained teamsters, so the tractors were making their way into more jobs.

I was in the farmhouse one day, making lunch, and lifted my head at the unusual sound of one horse on the driveway, being walked by one person, back toward the barn. Then another horse by another person. Then nothing. That was strange. I'd watched a three-horse hitch go to the field earlier, so two horses going back to the barn didn't make any sense. Also, if we had to walk horses somewhere, one person would lead two or three or even four together, the horses shoulder-to-shoulder on either side of the person. What was with the extra person? That was when my heart started pounding.

There's a difference between knowing in your head that a risk exists and feeling it in your body. The sound through the window of a situation going bad would cause my adrenal glands to send pure fight-or-flight energy into my bloodstream. Mothers who can't find their toddlers are familiar with this feeling. Those who have barely escaped a car accident. Usually, when I felt it, it turned out to be nothing—just an innocent shout from the field or horses in high spirits. Still, I was on guard when I knew there were horses at work. Part of my consciousness would always be aware of who was driving, which horses they were using, what they were pulling, and calculating the potential for trouble. On windy days or in spring, when the horses were spooky or not yet used to daily work, or with green teamsters or certain combinations of horses, I never relaxed until I knew they were finished and back in the barn.

My worst runaway happened in our third year. I had spent hundreds of hours working Sam and Silver by then, but I was still a green teamster, and I made some mistakes. We had been spreading compost in Monument Field. Spreading is brief, intense work that builds muscle on a team, like a couple of football players hitting the tackle sled over and over. The pull was easy on the way to the field, even with two thousand pounds of compost in the box, rolling over the hard farm road thanks to the magic of wheels. But once in the field, with the ground-driven web engaged, the horses had to work to push the compost backward as well as pull the cart forward, while the heavy box sank into the soft tilled earth. They needed all of their coordinated strength to make the first few dozen yards. As the compost was tossed out of the box and over the ground, the load became progressively lighter until, by the end of the field, they were pulling only me and an empty box.

The spreader, like most of our equipment in those early years, was cobbled together from antiques, some bought at auction, some pulled out of a neighbor's hedgerow. This one was unique—a horse-drawn spreader that had been converted to tractor, which we converted back to be pulled by horses. It had a tractor seat stuck on the front of it and, in place of a solid footboard, a flimsy piece of rusted metal set slightly too low for me to reach, so that my legs dangled like a child's.

Over the course of the week, the team had grown comfortable with the work. They liked the routine of it, the same thing over and over. Silver seemed to enjoy the opportunity to show off his muscles. I'd noticed, though, that something was irritating Silver. He tossed his head a lot, hung back, and needed encouragement to keep up with Sam. Maybe the bit was bothering him? That morning, I had switched his usual Liverpool bit, which had shanks, for a thick, mild snaffle, which offered no leverage between my hands and his mouth.

Near the end of my tenth load, I heard a disconcerting snap. I looked back to see that the compost was no longer coming out of the spreader. I stopped the horses and jumped off my seat to find that the ground-driven chain had broken and was dangling underneath the box. This was the price of old equipment. Something was always breaking, something so antiquated that the new part would almost always have to be improvised; we lost a lot of time that way in those years. I pulled back the lever to switch off the gears and rode the spreader back to the shop, where I called Mark to fix it. Fixing a broken link was a routine job, so we didn't bother to unhitch the horses, and I stayed on the seat. The horses were happy to have a break, and all three of us relaxed while Mark worked.

When the links were joined again, Mark told me to put the spreader in gear and move forward to test it. I did it lazily, without my full attention. The horses were half dozing when they began to walk forward. When I engaged the chain, the noise startled Silver, who lunged forward, a full neck in front of his partner. I felt a shot from my adrenal glands then, knew I had a quarter of a second to gain control of the situation. I pulled Silver in, but the mild bit, with no leverage, was nothing to him. I don't think he could even feel it. Because he was in front of Sam, I had no contact at all with Sam's mouth, the lines to his bit slack. In that instant, Sam leaped forward to keep up with Silver, and then they were both at a run, and I was focused only on staying perched on the precarious seat.

I tried to steer them west, into the wall of the well house, but didn't have enough strength to control their direction, let alone slow them. Instead, they turned east, through a section of the farmyard where we parked equipment we weren't using. I scrambled to find the flimsy footboard to brace my feet so I could use my arm strength on the lines. But it was more than enough just to keep my balance up there. I was aware that if I fell off, I could easily be run over by the heavy metal wheels.

Two horses hitched together and running away are a team of conjoined dancers in magic shoes, like the witch at the end of *Snow White*. Once they are going, they must go. If one falls, he will be dragged by his partner. They make their decisions independently, but their fate is tied together by thick leather and heavy chain. Sam and Silver snaked through the equipment yard at high speed, avoiding collision until Silver ran too close to a single-axle trailer resting on its tongue, the bed sloped up in the rear. Instead of stopping, Silver ran over it. The wagon tilted to its balance point under his weight, and he hesitated. In that moment, the spreader came to a sudden halt, and I was thrown from the seat. Somehow I landed on my feet, just in front of the spreader, and jumped clear. I was firmly on the ground, the lines still in my hands, dimly aware of Mark in the background yelling, "LET GO." But my brain was telling me to hold on. I was responsible for these horses, and if, by chance, they made it without lethal injury through this obstacle course of equipment, they were headed straight for a steep embankment and a ditch. I skidded alongside the spreader, leaning back, lines in my hands, pulling with all my might.

What happened next is preserved in my memory in narrow focus. The horses cut close to a stout sapling at the edge of the barnyard, just before the flat ground dropped off toward the ditch. I gave one last effort to pull them to a halt and then, too late, realized that I had cornered myself into a bad angle with the sapling. It happened so fast. The lines were running through my hands as the heavy metal wheel of the spreader hammered toward me and caught my left hand between it and the sapling before continuing on, the horses disappearing over the hill.

I was wearing a thick leather glove, and in the shock that followed, I could see the tracks the wheel had etched in it, and that was how I knew I was hurt. I yanked the glove off just before swelling made removing it impossible. The glove had saved most of the skin, but there was a tear over my first knuckle through which I could see the white glint of the bone.

Mark came running around the corner. I showed him that I was not mortally injured, and with the clarity and bossiness of shock, I told him to take care of the horses, who had disappeared down the embankment, still towing the spreader. I walked inside and dialed our neighbor Beth, who drove me the thirty-five miles to the hospital. The stoplight at the entrance to the Northway was red, but she looked both ways and drove through it. "Optional stop," she said casually, which was how I knew she was worried. I sat in the front seat and let myself feel the pain that was happening in my hand. It was something, that pain. Enough to stop chatter and demand my full attention. Beth got me through the emergency room and to a physician's assistant in the vain hope that all I needed was a little stitching. The PA said, "You need an X-ray." The radiologist said, "You need to see a plastic surgeon right now." The bone of my knuckle was broken three ways, and there was damage to the tissue around it. But considering how bad it could have been in that circumstance, I knew I was extremely lucky. There must have been half an inch between the wheel and the tree, which was probably the difference between having a finger and missing one. They gave me a humongous bolus of pills that blunted the pain and slowed my frantic thoughts about what might have happened to the horses. In my haze, with my good hand, I dialed the number at home, which rang to an empty house.

I didn't talk to Mark until I got home late that night, my hand set and bandaged. He had been doing my chores and had not heard the phone. I wasn't yet sure if those chores included burying two good horses. Finally, he came in and told me what had happened.

He had run to the embankment and looked over. The horses were there at the bottom, still in harness, with the spreader behind them. The horses were both down, mired in the mud of the ditch and tangled in the traces. Mark raced down and untangled them, cutting the straps he couldn't untwist, and took the horses by the bridles. First Silver. He rocked his large body once, twice, then stood square on all four legs and shook himself, as though he'd just had a good roll in the pasture. Then Sam. He got his feet under him and walked out of the ditch too, trembling slightly but completely sound. Mark led them back to the barn. Aside from a deep scratch on Silver's foreleg, they were unhurt.

So that day, when I heard the horses walking back from the field one by one, my heart started racing before I even got to the window to look. At first I saw nothing. The driveway was empty. Then I looked toward Pine Field to see, at the far edge of my vision, the machine that the three horses had been pulling—the pulver-mulcher—pinned in the trees and Mark trying furiously to lever it backward with a pole. There was a large chestnut-colored inert shape on the ground in the tangle of trees that could only be the third horse.

My mind flicked through the chestnut horses who might have been in the hitch that day: Cub, Abby, Jake, maybe Jay or Jack. I wanted reflexively to choose one, to make the horse on the ground the one it would hurt a little less to lose, as though I could make it so by wishing. I ran toward them down the driveway without feeling the ground. When I got close, I could see that the horse was wearing a bridle with a pink noseband. Abby. Our best mare, in the prime of her life, a brave and honest worker. Mark pulled off her collar and bridle and the harness too, cleared away the pine brush and the gear that had snapped, so as she was lying there on the ground, she was pure animal, just hair and hoof and sweat and breath, and not the civilized beast we made of her every day with all our straps, buckles, and training.

David had been driving the horses. He sometimes reminded me of Silver, our first gelding: 99 percent of the time, reliable, predictable. Because of that, I tended to ease into trusting him. But that crazy 1 percent! It was extremely dangerous. David was well-meaning, hardworking, kind and patient with children and visitors, and generally had good judgment. But 1 percent of the time, his judgment was skewed enough that if you combined it with bad luck and inexperience, it could get someone killed.

David had made a grave pair of mistakes. Either one alone could have caused a wreck; together, it was almost inevitable. He'd gotten off the seat of the pulver-mulcher and left his lines wrapped around it, then walked to the front of the horses to adjust a bridle that was rubbing Jake, the middle horse, the wrong way. The first rule for new teamsters is to never let go of your lines. When you need your hands to adjust machinery, or fasten a buckle, or open a gate, or anything, the lines are over your forearm, right line on top of left. Without lines in your hands, no matter the arrangement, you have no steering wheel, no brakes, nothing. You are a situation waiting for a complication.

David's second mistake had been removing the bridle from Jake while he was hitched. Not only is there no way to control the horse after that, but he is suddenly made fully conscious of the thing he is pulling behind him, which has been hidden from him all this time by blinders. Something like that had happened to me once while raking with Jay and Jack. I had left Jay's throat latch too loose, and he had managed to rub his bridle halfway off against Jack's neck. Jack's steadiness had saved us then. He'd planted his feet and stopped as I scrambled off the rake to run to Jay's head. But David was not as lucky. The sudden removal of the blinders, plus the unfettered freedom, was disaster.

The horse jerked up on his haunches, and David struggled to control them all for a few moments from in front, and then, thank goodness, he jumped clear. The horses were off, galloping together across the field with the pulver-mulcher clanging behind them. When they reached the hedgerow, still running, the outside horse—Abby—hit a young spruce tree with her neck yoke. The force of the impact slung her backward and down and stopped the other horses.

She was on her side with her belly against a tree when I got there. "She's alive," Mark said, which was the best he could say at that point. She couldn't rise, but we didn't know if it was because she was injured or because of the way she was pinned against the tree.

It is disconcerting to see a big horse on the

ground, in that position of unnatural vulnerability. Her flank was moving fast; her eye was large and looking right through us. What I felt most deeply was that we'd betrayed her. They lay their instincts at our feet when they are broke to work. In order to surrender their freedom, they trust us to keep them safe.

Mark and the two others took hold of her legs and, together, turned her over her backbone to her other side. If something were broken, this movement would make it worse, but if something were broken, she was done for anyway. She rested for a moment, breathing hard, then lifted her head, tucked her legs under her, and came to her feet, quivering. Mark walked her out of the patch of hedgerow, and we all watched, silent. Her legs worked. No lameness, not even a hint of how sore she would surely feel later. There were some small cuts and scratches on her belly and flank from the pine trees, but none of them deep. Her ribs moved in and out, and her breathing seemed even and easy enough to make us hope there were no internal injuries, no fractures. Mark clipped a line to her halter and walked her onto the farm road. She stood there for a moment, looked behind to the field she had crossed in uncontrollable flight, then dropped her head, shook her body, and walked back to the barn to rest.

David left the farm. Abby recovered without

complications, physical or mental. But I loosened my attachment to the horses after that. The farm was too big, the risk too great. I didn't think we could devote the time that Donn did to work with his young teamsters; nor could we simplify our farm to be more like theirs. I didn't think we could hold up our end of the deal with the horses and keep the farm going at its current scale. After that, we used them less and the tractors more. If someone with horse skills arrived to work at our place, we put them to good use. But we stopped training people to drive. Some of the horses still worked regularly, but others spent a lot of time on pasture. I pestered Mark to sell them, at least the teams we weren't using, but he didn't want to. He wanted to hold on to the link to see what came next.

Nothing remains. Not the bottomless heat of the sun nor your loved ones' health nor the lubricity of your joints, your youth. Maddening impermanence. At middle age, that hot grief stokes the market for sports cars and Botox. The children growing up. The parents growing old. Summer was beginning to fade, all fin de siècle abundance, entering its age of decadence.

Mark and I walked past the overgrown lawn in front of the farmhouse, where goldenrod bloomed. Some milkweeds had escaped grazing and mowing, and into their late blossoms, the monarch butterflies feasted. A lawn is a sentimental vestige of an idealized pasture, the extrapolation of the idea of an English country estate, populated with decorative sheep. My own lawn had grown wild and coarse, beyond anything a sheep would ever want to eat. It was grazed by butterflies instead.

The hens were pastured in Superjoy. One of them was outside the fence. She fixed us with one small black eye, her head bobbing over a wave of long grass. There is nothing so foolish as the sight of a grown person trying to catch a hen. In an open field, the hen always wins. But Mary saw her too and looked to me for permission to apprehend. "Sh sh," I said in encouragement. She launched, faster and more agile than I and smarter than the hen. She stepped on a wing to hold the bird and then pinned her with her mouth, over the neck. She didn't chomp. She held. The hen submitted, as if to a rooster, squatting suggestively, suddenly still, with her wings tucked in a V behind her. I walked toward them, told Mary she was a good dog, and picked up the hen, a neat package of feathers. I tossed her over the electric net, and she rejoined her sisters, who were squabbling over a dead toad.

While Mark went to inspect the cabbage, I sat outside the chicken fence and watched. Spectating and not doing is a form of indulgence we call farmer TV. There is not always time for it. I could have been bedding the nest boxes with fresh straw or cleaning out the water basins. But sometimes we indulge ourselves. This particular show should have carried a trigger warning. Watching the hens with a toad reminded me of middle school girls. Innocuous in the singular but, as a group, bloodthirsty, brutal. The last thing that poor toad saw was a bunch of feathered giants converging on him, beaks raised. Chickens are omnivores. They eat bugs, mice, pretty much anything they can get their beaks into. The toad was a coveted piece of protein. They all wanted it, and none of them could eat it, because in order to do so, the hen with possession would have to drop the toad, aim, and peck. But she couldn't even get to the dropping part, because every other hen on the field was trying to sack her. The toad went from hen to hen, the whole flock at full speed, to the soundtrack of excited caws and cackles. When Mark returned from the cabbages, they were still at it. I wondered how it would end. Maybe with the sunset. For hens, the world ends every night. As evening comes on, they sidle toward their roosts. They trudge up the ramp, settling in, and they are still until the sun comes up again. A hen caught by darkness away from her roost is, well, a sitting duck, as good as paralyzed until the light returns.

CHAPTER 17

I'm still working on separating my ego from outcome and my own happiness from Mark's. Focusing on the joy that comes from working hard every day and not on its result.

My friend John is a painter. He just turned sixty. He arranged his life carefully from an early age to make it suit his art, which he has practiced every day without fail. Sometimes his work goes well, and sometimes it doesn't. Recently, he told me, he walked into his studio and realized everything in it was heading in the wrong direction. He stood in front of a painting he had worked on for months and began again. "Sometimes you kill the shark," he said, shrugging, "and sometimes the shark kills you." But you don't ever stop swimming. John has been going so deep into his work all these years that everything else disappears. When he's in it, nothing can touch him.

A few years ago, he fell in love with a Frenchwoman, and they got married. She left him suddenly, soon after her immigration papers were arranged, and she took a lot of his money. He was ruffled for a while but basically unfazed. The thing that is most deeply him—his soul, his work—can't be taken away.

Mark is like John, with the farm. He walked into the work of it every day, with intensity, without fail, until he *became* it. He works with acquired detachment from any individual success or failure, his eye on the larger picture. I don't have John's steadiness or certainty. I don't have Mark's focus. In my work, I vacillate wildly between shark hunter and chum. But I know both Mark and John are onto something good that grants peace and happiness. The farm taught me that. When you look at a farm from the outside, it looks like work is the cost. From the inside, you find that the work is the reward or, rather, the work is all there is, and it's a beautiful thing.

The Amish see it more literally. To them, work is a form of worship, a means of glorifying God. They sing at work, an ethereal, strange, and dissonant sound. How do I know that? Because just as the wave of young people interested in farming slowed, the Amish arrived in our region. They settled the old pieces of land around us that had fallen into disuse. They brought skilled hands and a desire for work. It's a good thing we held on to those horses. We have Amish people who work at our farm now, and the horses are back in the fields.

When Racey's son was ten months old, there was a fire. I heard the call come over the pager as I was getting ready for bed. Her address, structure fire, fully involved, the pager said. That's a very bad string of words. It didn't say which structure. I thought of her with her baby in the basement of the new house and felt a chill. I sent her a text, and she texted right back, surprised to hear there was a fire at her farm. She was safe inside with the baby, but when she got my text, she looked out the window to see their new barn in flames. Nathan had been nursing a sick calf out there, had left a heat lamp on. Maybe it was the heat lamp, maybe it was bad wiring. Maybe it was the brooder that was in the other corner, full of chicks. Nobody knew. Someone had seen it from the road and called the fire department. The neighbors were there in seconds and, in another minute, the truck. By then Nathan had seen the fire and come running, and so had Chad.

But it was far too late. The whole building was engulfed, burning fast, hot, and bright. There were no horses in the barn, just the calf, the chicks, and some laying hens. It burned to the ground, along with all their tools, the combine, the hammer mill, fencing equipment, and a year's worth of hay and grain.

The next day, I made a potpie to take to them. What else is there to do when your friends are hurting but make them food? That's the thing about misfortune and grief: they have a lot of weight but no mass. You can't lift them on behalf of a friend. But at least you can bring food. There is something inherently comforting about a potpie. The savory sauce, the rich chicken and mushrooms, all covered by a blanket of pastry. Smothery love in a dish.

When I pulled into the driveway, the shock of it hit me. The barn that they had worked so hard to build was gone. I got out of the car bearing my potpie and burst into tears. It was sorrow not for the building, which was just a building, but for the loss of the work that went into it. The building and everything in it could be replaced, but the work, on a new farm, is imbued with all the hope for the future, not yet sullied by bad harvest, disappointment, quarrels. I cried for the setback and the burning up of hope. When I saw Nathan and Racey, though, they were resilient, strong. They were glad their family was safe, sad for the loss of the barn, but ready to build again.

Mark and I went back after the insurance inspectors had left. It is something to see the aftermath of such a fire. It's not the absence of the whole that gets you but the odd parts that remain. In one corner, there was a set of intricate gears, teeth still perfectly aligned but with the case of whatever had contained them burned away. In another corner, chickens roasted in their feathers, like a demonic barbecue. A few exploded rats, who must have been very rudely expelled from their cozy home in the grain or under the hay. Around the perimeter, there lay the almost-new hardware that held the barn together, only now without the wood, which had turned to smoke and ash.

They rebuilt, of course, and finished their house. They had another baby, a girl, and built the farm up, one field at a time. Chad and Gwen built their new home across the street and got married there. They share work and equipment with Racey and Nathan. Chad's Suffolk stallion has sired foals all over the region. Blaine and Tobias got married too, and built a beautiful farm together in a town nearby, where they raise beef, pork, and lamb and run a custom butcher shop. They have a new baby, a little girl. Tim's farm is farther away, south of us. He has a CSA that produces vegetables and eggs for his year-round members and makes good use of his team of Percherons. We all see each other as regularly as our work permits, to share advice or tools, or trade produce, and talk as equals about the challenges of pulling a living out of the dirt.

Our farmhouse, though, was still a problem. The thought of returning to the dim and dingy house every day continued to depress me. According to Mark, it was never the right time to renovate. All years, all seasons, all weeks and days were full. Every dollar that came in was quickly whisked into the large needy maw of the farm. Our assets increased a bit, but not our comfort level.

After living with Mark for a decade, I'd come to understand that he felt true and deep-rooted antipathy toward the idea of fixing anything in the house. On the farm, no project was too big, no ambition too grand; the house was where he stowed all his hesitation and fear. If I brought up a small improvement, he listed the things that could go wrong, the ways it could balloon into some crushing job that sucked away all our time and money. Raising the topic of house renovations pretty much guaranteed a fight. Which didn't always stop me from raising it.

But each of us is responsible for our own measure of joy, and this frustration with our living space was sapping mine. Eventually, I called Candace to talk us through it. She used her powers to make us look at each other and explore our differences honestly, calmly, and with empathy. We dug down, painfully, until we got to the bottom of it. He was afraid that once we started tearing things apart, I'd just want more, the cost would be overwhelming, and we'd live in a construction zone for eternity. I was afraid we weren't important enough to him, that he cared about the well-being of the farm more than that of me and the girls. By the end of the session, we'd agreed on two things: a small budget, and that the entire project was up to me. He wasn't going to stop me, but neither was he going to help. After that meeting, I stood in the cramped dining room, the dreary playroom, feeling a little nervous. I didn't know anything about renovating a house. I had no idea how I was going to get it done, and now that I had his grudging consent, I was afraid to start. For a long time, nothing happened.

Then, just after Christmas, Mark took the girls to Montreal for a few days' vacation in a gorgeous penthouse apartment with a panoramic view of the old city and a private skating rink on the ground floor. A friend had loaned the apartment to him. This is what my friend Alexis calls fancy squatting. Mark *adores* fancy squatting, because it combines luxury and thrift; somehow he ends up in a lot of fancy squatting situations. I packed the girls' bags and kissed them all goodbye, wished them a happy New Year, and stayed behind to take care of the farm and catch up on paperwork in the blissfully quiet house.

Early on the second day of the New Year, two Amish teenagers knocked on the door, looking for Mark. It was Do What You Will Day in the Amish community. I knew about this tradition. When I'd first heard about it, I thought it meant they could do whatever they wanted, but this was a very conservative Old Order group, and it wasn't quite that good. The day after major holidays—Christmas, the New Year, a half-day after Thanksgiving—the young people are allowed to look for work wherever they choose, and to keep their pay instead of contributing it to their family, as they usually did. The boys had driven their horse an hour to our house in their black wool cloaks. I could see the buggy and the horse tied to the pole barn, a blanket thrown over his steaming back. They wanted to know if Mark had any work for them for the day. It was a good bet. If Mark had been there, he certainly would have. He always had a to-do list in his head. But I couldn't think of a one-day two-person job that wouldn't require more supervision from me than it would be worth. "I'm so sorry, he's not here," I said. They started to turn away. "Wait!" I yelled after them. "Do you know how to tear down a wall?"

They unhitched their buggy and put the horse in the barn, then went to the machine shop to look for sledgehammers and pry bars while I ran to the basement to shut off the circuit breakers. *Wham! Wham! Wham!* The bars bit through a wall and its leprous layers of wallpaper. *Wham!* A few more hits, and the light shone through, and the two small rooms were on their way to being one big one. Soon the whole house was a fog of plaster, splintered wood, and hundred-year-old dust that settled onto the roaring woodstove, sending up a scorched smell that filled me with joy. The Amish kids threw open the windows and chucked the old wall outside in pieces, which I shoveled into the bed of the truck and drove to the Dumpster. Then they ripped down the horrid acoustic tiles on the drop ceiling. Those fell to the floor, with their coating of grease and dirt. Then down came the damaged old horsehair plaster and lath ceiling underneath. They shoveled everything out through the window. Mark called from Montreal in the middle of it to tell me about the skating rink. "What's all that racket?" he asked. By the time the buggy disappeared down the driveway late that evening, I was white-haired with dust and bone-tired, and the downstairs of our claustrophobic house was laid open, its depressing old elements gone.

It took a year and the help of several friends to finish all the details, but when it was done, it was perfect. The dingy maple floors that had made me so sad to look at were clean and smooth, finished with a linseed-based hard wax that I put on myself, which turned them the color of new honey. The molting wallpaper was gone from the old plaster walls. The cracked too-small windows were gone too, new ones fitted into the original framing, so the house no longer looked like it was squinting, and the room was flooded with light. Ronnie's husband, Don, copied the intricate original molding so faithfully in his woodshop that you would never know they were new. My designer friend Mark Hall figured out a clever, simple way to punch through two walls and access a staircase that adjoined the main floor with the bedrooms so that at last, we didn't have to go outside in order to go to bed. His wife, Erin, who is also a designer, helped me pick out a warm white paint for the walls, a shade that makes me happy in every light. The budget ran out before I got the ceiling replastered, so I left it open to the rough rafters. It was fine. It was perfect. I liked seeing the old house's good bones. Mark tied two swings to the rafters so the girls could sail through the air the whole long length of the room. In the winter, the heat of the woodstove seeps through the floorboards and rises up the stairs so our bedrooms are no longer frigid. I made sure we installed enough warm, soft lighting that it would never feel dim or dingy again.

We didn't need a new house after all. When something isn't working, you don't always need to tear down the whole structure. Sometimes you just pull down a wall, reinvent, put some new skin on the old bones. And it's lovely.

EPILOGUE

There was a cold snap, the temperature dipping into the mid-teens. In the morning, we skated on the pond, all four of us. The ice was over an inch thick but transparent as still water, unscathed until we ran across it with our blades. Underneath, at the edges, we could see the crayfish and water bugs eyeballing us from the other side. The wind was so strong we reached our arms out and let it push us across the surface. I skated circles around the edge, and Mary ran laps on the bank, barking with something between joy and frustration. She couldn't understand the sudden solidity of the pond, and she wanted very badly to work.

Mark and I took a farm walk. After frost, and before the monotone of winter, the farm was in a state of transitional beauty, at the threshold between life and death. The tomatoes on the ground were bright red and softening into the earth. The raspberries clung to the canes, but their life was gone, their color washed from vibrant acrylic to soft gouache. Only the hardy chickweed remained green over frost-rimmed ground. It had taken hold around the tomato stakes and spread from there. It was tenacious, low-growing, and strongly rooted, a formidable enemy. It would give us trouble in the spring.

"It might be time to mow the asparagus," Mark said as we pushed our way through fronds that towered over my head. In the summer, the girls had hidden in them, and woven them into fairy crowns and wide green skirts. As the days had grown shorter and frosts came and went, the green slowly drained from the tops of the plants downward, until the whole plant was brown. That was how you knew the plant's energy had left the surface and gone underground until spring. That was an image I liked, the life force of the plant disappearing from view but not from existence, returning to its source, and waiting there until the sun called it to incarnation in the spring, its energy—that notion thin as a ghost—waiting to knit itself together into being again, out of the threads of carbon and nitrogen, water and minerals.

But there really is no clear line between green and brown. As with all things in farming, there is no simple answer. Most of the asparagus stems were brown but not all of them. To preserve the plants' energy, it would be best to wait a few more weeks before moving. But meanwhile, the asparagus beetles, the specialized pests that plagued the plants, were on the march, climbing down the stems, aiming to burrow into shelter for the winter. We had to balance the plants' needs with our desire to kill the beetles. It's always a choice between imperfect sides. We decided to mow as soon as possible.

Mary came on the walk with us, and Jet trundled along at our heels. When Mary arrived, he relinquished all his responsibilities to her. He seemed content in his dissipation. Mary ran circles through the cornstalks and then bounded back to him, took his snout into her mouth, draped her forepaws over his broad back. He flapped his tail slowly back and forth under her assault, like an old veteran waving a flag at the edge of a parade. She was manic circles of youthful energy; he was mature conservation of effort.

The farm looked fecund, even in the mode of early decay. The cover crop was particularly beautiful. There was an impressive stand of oats, peas, and tillage radish planted to add carbon, nitrogen, and friability to the soil. The radishes had dug six inches into the topsoil, their taproot three feet. When dead and rotten, they would leave organic matter and airspace behind them, giving next year's plants the underground oxygen they needed to thrive. Mark pulled up a radish and took a bite. Tillage radish is a sibling to the daikon, and it looked like a long, white, thick carrot. The flesh was frozen at the top but fresh and sound at the bottom, where the frost had not yet reached. I took a bite. It was mild and good. The peas, most sensitive to frost, were shriveled brown tendrils, twelve inches long. The oats were still green. Mark swept up a big handful and chewed it, so that for the rest of the walk, the corners of his mouth were stained green from the juice.

Mark is a person whose intensity and sheer capacity for joy make him want to consume the things he loves until their essence drips from his mouth. It's a lucky thing, really, that his vices are food, farms, and work.

The hot peppers were hanging dead and limp on the vines. "We could harvest these," Mark said. That's how chili powder is made, and dried pepper flakes. There were a million things exactly like that in the fall. Things we could do or should do, things we wanted to do, but there simply was not time or bandwidth to do them. A part of me believed there should be time, and there would be time, if only I were a better person—more efficient, less messy. I was old enough to know, deep down, that even if that were true, it was not possible to will myself into an entirely new shape, different from what I have been.

The last sun threw our shadows in front of us. Even all these years in, I find the difference in our height startling. Until I see us reflected somewhere, I forget I am small and he is so tall. We seem like different species. Mark looks older than when I met him. His face and neck have been creased and permanently colored by the sun, and his sandy hair has crept backward a little, grown thinner. In the unfair way that some men age, it all looks good on him. He is still lean and strong, and his familiar hand is warm and callused. His physical presence is a pleasure to me.

The moon came up, gigantic and perfectly full. I had heard on the radio that this was the month of the supermoon, when the two unlike bodies—one large, one small—are as close as they ever get, and the apparent size of the moon is at its most glorious. *This is my place in the world,* I thought. *Five hundred acres of soil, between mountains and a lake.* I was a foreigner here not that long ago. But I have eaten what came from this soil and the rain that fell on it until my cells were made of it. This was the raw material that knitted two children together in my body. And if a person's essence is made not of atoms but of thoughts or of time, still, by any measure, I am made of it, and I belong here.

The girls played outside in the dark until dinner, in the half-frozen mud. For dinner, we had squash soup, good bread, kale. The soup was from a roasted, pureed blue Hubbard squash, thinned with a Chinese master stock, and scented with anise, ginger, garlic, cinnamon, soy, and black cardamom.

After dinner, we turned on They Might Be Giants, and the four of us sorted beans that needed to be packed for the city shares the next day. The beans we'd grown were called King of the Early; they were fat, mottled, wine-red and creamy-white. The fire crackled in the woodstove. We talked, laughed, and worked. We started with a great heap in the middle, the four of us sorting on four sides of the table. Jane turned out to be the fastest and most accurate sorter, her nimble eight-year-old fingers picking out chaff, broken beans, and stones. When the central heap was gone, her pile of perfectly sorted beans was bigger than Mark's and mine put together.

It struck me, that night, that it had gotten easier. Something shifted after Miranda turned five, just as Candace had predicted. I didn't feel that clutching in my chest anymore if I didn't know where they were every second. I didn't worry, when I ran to the barn to check on a new calf, that I would come back to the house to find one in tears and the other bleeding, or the house on fire with them trapped in it, or any of the other irrational instinctual things I once feared. I had the sense that they were weaned again, in a different way. I could have back the piece of myself that they had needed for those years, and they could begin their exploration of the world away from me. It was a relief, tinged with sadness because there was such great love and satisfaction in the exchange of need, and it would never come back, at least not in the same way.

It wasn't just parenting that got easier. Marriage got smoother, Mark got healthier, the farm more stable financially. We learned the art of being grateful for what is, instead of longing for what isn't, and the benefits of separating our own well-being from that of the farm, and even that of each other. As we slid through our forties, we began to settle into the prickly truth of middle age, which at its best is a time of acceptance, self- and otherwise. The hardest piece was something the farm was very good at teaching: Everything ends. The natural order of things is immutable. Seed, flower, fruit, decline, death, decay. Seed. Each stage has its own drama and its own particular beauty. If you can see it, you can accept it. The parts are graceful, and so is the whole.

ACKNOWLEDGMENTS

Thanks to my agent, Flip Brophy, for being my advocate and friend, and to Nell Pierce for all her help. Heartfelt thanks to my editor, Kara Watson. Her combination of skill and kindness made this a better book by far, and the process of writing it so much more enjoyable. I'm grateful to Beth Thomas for excellent copyediting, to Emily Greenwald for her stellar assistance, and to Nan Graham for giving me a wonderful home at Scribner. Thanks to Judy Goldschmidt and Nina Nowak for being first readers back when this was a big messy pile of pages, and for friendship and support when I needed it most.

One of the hardest parts of getting a book out the door is finding a quiet place to do it. The Blue Mountain Center provided a much-needed haven. The Essex Volunteer Fire Department and Lakeside School offered a desk at any hour. Barbara Kunzi loaned me her whole house for the long final stretch. I am so grateful to her for that generosity.

It took a lot of help to keep the family and farm afloat while I was working elsewhere. Thanks to everyone who has farmed here with us and especially those who lent me their stories. I deeply appreciate your willingness to do it.

Barbara, Beth, Don, and Ronnie, thanks for nurturing our girls.

Thanks to my parents, Tony and Linda, for being my biggest fans always, and to my sister, Kelly, for so darn many things, but most especially for showing me what love and courage look like when deployed in equal measure. Finally, thanks and all my love to Mark, Jane, and Miranda. Life with the three of you is the greatest adventure.

Center Point Large Print
600 Brooks Road / PO Box 1
Thorndike, ME 04986-0001 USA

(207) 568-3717

US & Canada:
1 800 929-9108
www.centerpointlargeprint.com